HIGH CARBON MANGANESE AUSTENITIC STEELS

F. Maratray, *Consultant*

Edited by

THE INTERNATIONAL MANGANESE INSTITUTE

17, avenue Hoche
75008 Paris
France

Maquette : J.-G. Vizet, M. Drouhin pour SIRPE éditeur, Paris (4e)

ISBN 2-901109-07-1

Dépôt légal : 2ème trimestre 1995

Foreword

High manganese austenitic wear resistant steel was the first alloy steel ever produced. Invented by Hadfield in 1882 this steel had many applications due to its very high resistance to both impact and abrasion and also because of its high workhardening capacity.

Among these applications the best known are the parts used for different types of crushers in the mining industry, parts for earth moving equipment and parts for railways such as railroad crossings.

High manganese austenitic steels have properties which depend not only on their chemical composition but also on the way they are produced and processed.

It is very difficult to develop a general theory applying this steel grade as such, therefore, this publication is not aiming to be a critical review of the literature by supporting a physico metallurgical presentation of the Hadfield steel. Many results found in papers already published conform to the specific sample of steel used for the tests and can only be extrapolated with care.

Mr. Maratray who was in charge of the important task of reviewing the literature tried to select information and data of interest to both the producers and the users of this wear resistant steel.

The presentation was combined with as much practical data as possible which are given in tables or illustrated as figures.

Manganese austenitic wear resistant steels were examined from a scientific and a practical point of view.

Some missing data was obtained from tests or calculations specially selected for this purpose.

We wish to thank : Creusot-Loire Industrie, Division Marrel - Le Creusot et Usines et Acieries de Sambre et Meuse S.A., Usine de St. Brieuc for the supply of test samples of Manganese Austenitic Steel.

We would like to express our thanks for the work Mr. Maratray has accomplished and hope that this publication will be of considerable interest to a wide range of readers who are interested in the production or the utilisation of wear resistant manganese steel.

We also wish to thank Ms. L. Cooper who assumed responsibility for all the secretarial work.

P.L. DANCOISNE

High-Carbon Manganese Austenitic Steels

Historical Background

The first manganese steels were produced by the Terre Noire Company, where Alexandre Pourcel had perfected the manufacture of a ferromanganese in the blast furnace, at a reasonable cost. At the Universal Exhibition in Paris in 1878, the Terre Noire Company displayed several steels with different manganese contents, of which the most highly alloyed contained 2.4% manganese and 0.6% carbon.

R.A. Hadfield became interested in this work and he undertook a more detailed study of manganese steels. To this end, in 1878 he prepared a series of steels, adding ferromanganese in varying proportions to a decarburized steel, thus obtaining steels which had carbon contents about a tenth of their manganese contents. In 1882, he arrived at the conclusion that steels containing between 2.5 and 7.5% of manganese, water-quenched at 1 000°C, were hard and brittle, whilst steels with more than 10% manganese had a remarkable toughness. In 1883 he took out a patent for steels containing 7-20% manganese.

The main manufacturing difficulties of cast and rolled products were overcome in 1887, and samples were given to experts to determine their physical and mechanical characteristics. In 1903 F. Osmond showed that at room temperature Hadfield steel retained the stable structure of high temperature, and gave it the name of Austenite, in honour of the metallurgist Roberts Austen.

Before the First World War, the first austenitic steels were rapidly used in industrial applications for the manufacture of wear parts of grinding machinery, earth moving equipment, and in railways.

The standard grade of 1.2% carbon and 12% manganese proposed by Hadfield is still widely used. Other grades, more or less rich in carbon and manganese, with the addition of alloying elements provide a better compromise, in certain applications, between abrasion resistance and toughness. The metastable austenitic steels introduced by T. Norman at the end of the 1950's, were created with this intention.

Manganese austenites have other interesting properties which have enabled them to be used for industrial applications in other sectors. The use of manganese austenitic steels as a non-magnetic material came into being for alternator reinforcing rings and military applications. New grades with higher yield strength, and with a better weldability and machinability, were studied at the end of the 1970's for research equipment in nuclear physics, and for magnetic levitation vehicules and linear motors. The expansion coefficient of manganese austenitic steels can be reduced to that of carbon steels, thus giving rise to the possibility of these steels being used in cryogenic applications.

Large additions of chromium or aluminium give excellent oxidation and corrosion resistance to manganese austenites: chromium-manganese austenites are used for their resistance both to abrasion and corrosion, in exhaust valves and as hot-working tools ; manganese-aluminium austenites have excellent resistance to corrosion in chloride solutions, as well as to cavitation, and they are used for boat propellers.

The successful use of Hadfield steel for wear parts is a result of two properties : superficial hardening following work-hardening, and its toughness. The work-hardening capacity of Hadfield steel remained a mystery for a long time, and is still not completely explained. Deformation in a system of uniaxial traction, deformation or compression brings into play several mechanisms, of which the relative importance and interaction are not well understood. Explanations based on uniaxial forces cannot be applied to the surface deformation seen in service where steels have a considerably greater hardness than after deformation in strain or compression, where the deformation is limited by the occurrence of fracture.

Manganese austenites are very susceptible to intergranular embrittlement phenomena, which are the cause of most manufacturing rejects and service failures. The effect of mass plays an important role in dendritic segregation, the concentration of elements in inter-dendritic spaces acting directly on the toughness, and indirectly modifying the reaction to heat treatment, i.e., the kinetics of transformation of the austenite.

Laboratory studies of manganese austenitic steels have most often been carried out on samples taken from thin section ingots, and these properties are not easily transferable to industrial manufacture, as the results are not representative of properties obtained in heavier sections.

This brochure brings together data which can be of help in overcoming difficulties in processing and use for both producers and users of manganese austenitic steels.

Part 1 - Physical Metallurgy

I. *The Fe-Mn-C system*

II. *Work-hardening Mechanisms*

III. *Fracture Mechanisms*

IV. *Embrittlement Mechanisms*

I. *The Fe-Mn-C system*

a) Fe-Mn-C System

The Fe-Mn-C diagram was calculated by Thermodata(l) from experimental results published in the technical literature. The different phases taken

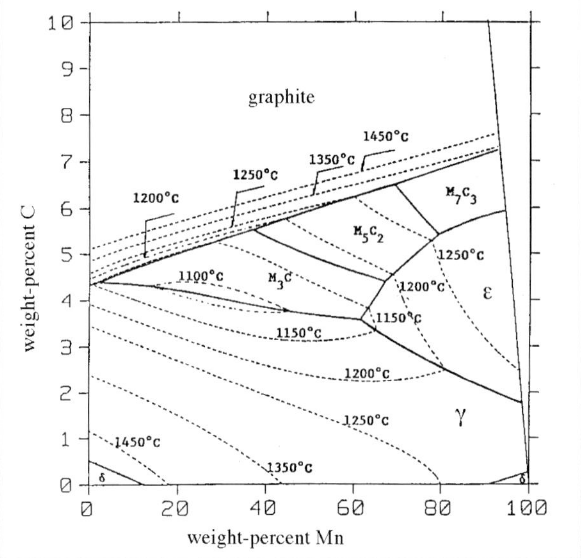

Figure 1 : Liquidus Projection in the Zone C<10%wt., Mn<100%wt (1)

into account in this calculation are the liquid phase, the carbides M_3C, M_5C_2, M_7C_3, $M_{23}C_6$, M_4C_{1+x} (ε), the graphite and solid solutions δ, α, and γ. The crystal structure of the carbides are given in table 1. The liquidus curve between 0 and 100% manganese and 0 and 10% carbon (figure 1) differs perceptibly from Schurmann and Giessler's(2) diagram. For more details on previous studies of the Fe-Mn-C system, see Rivlin(3).

The stability domain of the γ phase extends without a break from the Fe-C system to the Mn-C and Fe-Mn systems, where it meets the βMn_ε phase in the Fe-Mn system., around 60% manganese content, and carbon contents of less than 1% in the Mn-C system (figures 2-3). The chemical composition of the alloys which are of interest for their industrial applications comprises between 6 and 30% Mn and 0.4 to 1.5% carbon. The pseudo-binary cross-sections of 8-12 and 18% manganese (figure 4, p. 10) show the equilibrium area of the γ phase. The range of solidification temperature of the alloys under consideration is about 50% greater, for the same carbon content, than that for carbon steels, so they have a strong tendency to segregation and microshrinkage formation. The temperature and the carbon content of the eutectoid composition both increase when the manganese content decreases. The eutectoid point of an austenite at 18% manganese is at 0.23% carbon and at 544°C, and for an austenite at 6% manganese it is at 0.56% carbon and at 676°C.

The diagram of the stability domain of the γ phase at room temperature and at -196°C, established by Nikolin and colleagues (figure 5), was chosen in preference to that of other authors(8)(9) for its more rigorous experimental conditions (water-quenched at 1100°C as against 950-1100°C). The effect of carbon on the stability of austenite is very marked, an

Formula	Structure	Parameters (μm)				Observations
		a	b	c	β	
M_4C_{1+x}(ε) (x=-0.3 to +0.06)	Hexagonal	0.2709		0.4429		For 5.5% C, large composition area, the lattice decreases with decrease in C %
$M_{23}C_6$	Cubic D8	1.0585				
Mn_3C	Orthorhombic	0.5086	0.4573	0.6772		
Fe_3C	"	0.4525	0.5089	0.6743		
$(Fe_{1.2}Mn_{1.8})C$	"	0.5084	0.4528	0.6763		
$(Fe_{1.8}Mn_{1.2})C$	"	0.5074	0.4520	0.6757		
$(Fe_{24}Mn_{0.6})C$	"	0.5062	0.4511	0.6750		
M_5C_2	Monoclinic	0.5086	0.4573	1.166	97° 75'	
Fe_5C_2	"	0.5058	0.4573	1.156	97° 44'	Hägg Carbide
$(Fe_{1.1}Mn_{3.9})C_2$	"	0.5078	0.450	1.164		
Mn_7C_3	Orthorhombic	0.6959	1.1976	0.4546		Hexagonal symmetry is also proposed

Table 1 : Crystal Structure and Lattice Spacing of Carbides in the Fe-Mn-C System

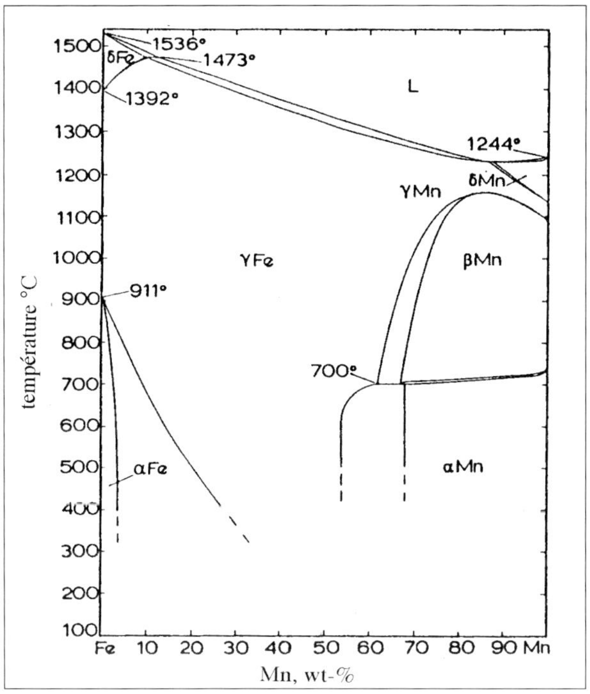

Figure 2 : Fe-Mn Binary System (4)

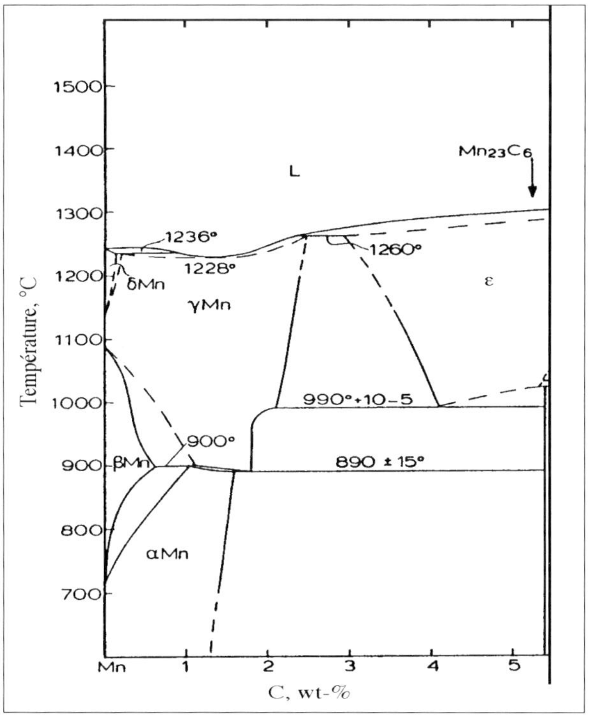

Figure 3 : C-Mn Binary System (5)(6)

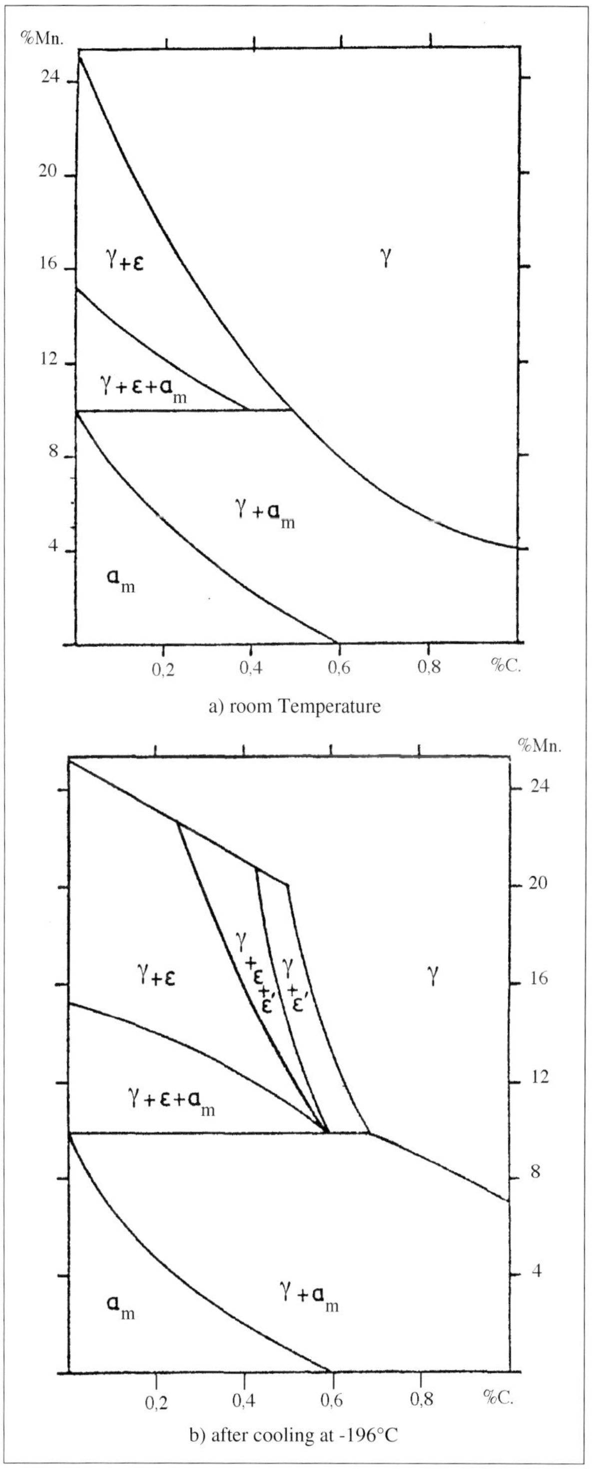

Figure 5 : Stability Domain of the γ phase in the As-Quenched Conditions (Water from 1100°C) (7)

addition of 0.4% carbon being sufficient to stabilize, at room temperature, an austenite of 12% manganese water-quenched at 1100°C. ε-martensite occurs in austenites containing more than 10% manganese and less than 0.5°% carbon, and α-martensite occurs in austenites with a lower manganese content and a higher carbon content.

The effect of carbon and manganese contents on the temperature of M_s points or ε_s of metastable austenites above room temperature, is shown in figure 6, and table 2 indicates the temperature of Ms point of several austenites which undergo the transformation $\gamma \rightarrow \alpha_m$ on cooling.

The domain of stability of the austenite decreases after cooling at -196°C in favour of the expansion of domains α and ε, and an intermediary phase ε' occurs between phases γ and ε.

The carbide-forming elements, chromium, vanadium, niobium and titanium lower the solubility of carbon

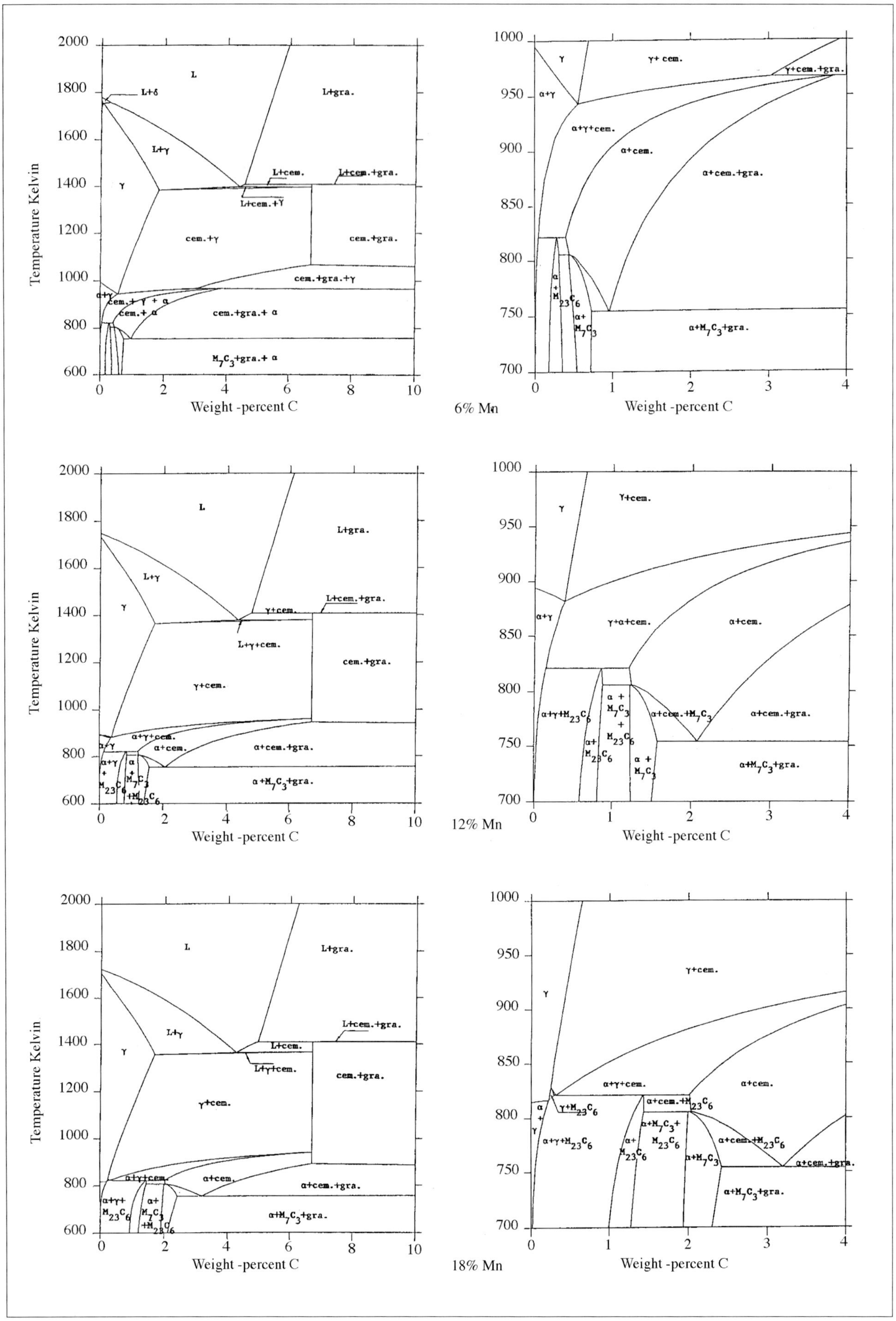

Figure 4 : Pseudo-Binary Diagrams of the Fe-Mn-C System at 6-12 and 18% Mn (1)

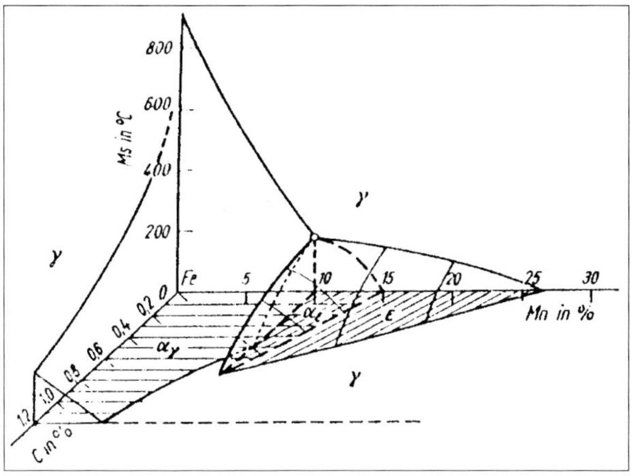

Figure 6 : M_S Temperature of Metastable Manganese Austenite above Room Temperature (8)

in austenite. For the same carbon content, the temperature of A_{cm} point increases after the addition of one of these elements. Industrial experience shows that to ensure the dissolution of the carbides in the austenite of Hadfield steels with 2% chromium, a solution treatment at a temperature of 50-100°C higher than that for standard Hadfield steels, is necessary, and a heat treatment of 2 hours at 1150°C ensures only a partial solution of carbides in the austenite of a steel with 0.8%C - 13% Mn - 1%V (12).

C %	Mn %	Si %	M_S (°C)
1.10	4.79	0.29	-19
1.03	5.71	0.31	-73
1.14	13.4	0.43	<<-196
1.16	7.75	0.33	<-196
1.16	6.03	0.29	≅-196
1.16	4.98	0.31	-57
1.15	3.97	0.29	-31
0.69	8.30	0.22	-2

Table 2 : M_S Temperature Values of Manganese Austenite (10) (11)

Aluminium prevents the transformation $\gamma \rightarrow \varepsilon$ in alloys of 15-30% Mn, and promotes the formation of δ-ferrite. The volume fraction of δ-ferrite is given as a function of chemical composition by the following equation :

$$1 < \delta = 33 + 2.6Al\% + 5.4Si\% - 1.6Mn\% - 8.5C\% - 1.2Ni\% - 4.6Cr\% < 8$$

This equation was established for casting temperatures between 1400 and 1450°C, and could be considered as valid in the composition range :

Al = 6 to 13%, Mn = 7 to 34%, C = 0.2 to 2.4%, Si = 0.4 to 1.3%, Cr = 0 to 6%, Ni = 0.5 to 6% (13)

b) Effect of chemical composition on segregation

Austenitic manganese steels used in industry (table 3) all have a hypereutectoid composition. They are situated, after cooling to -196°C, in the γ field of Nikolin's diagram, with the exception of the grades at 0.5% C, which are stable at room temperature, but which undergo the transformations $\gamma \rightarrow \gamma + \varepsilon'$ or $\gamma \rightarrow \gamma + \varepsilon' + \varepsilon$ at -196°C.

The solidification interval of these steels increases from 90°C to about 220°C, when the carbon content increases from 0.5 to 1.3%, and the segregation of the carbon increases in consequence. Carbon is not the only element which undergoes segregation : a marked segregation of phosphorus (figure 7), chromium and molybdenum in particular can be noted. The segregation of these elements is the reason for the different intergranular processes of hot cracking which are the cause of numerous manufacturing defects : amongst these are the effect of phosphorus atoms as impurities in the grain boundaries, of intergranular phosphides and carbides, and of low melting point eutectic cells.

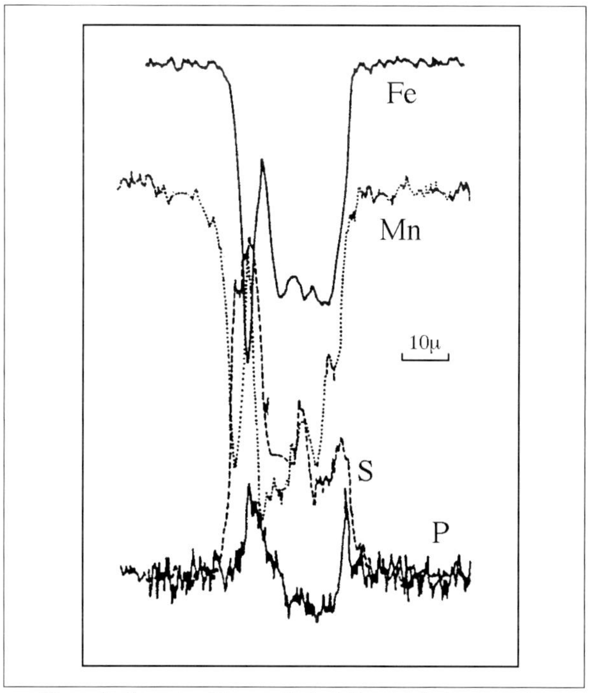

Figure 7 : Scanning Electron Probe Micro-analysis across the Austenistic Grain Boundaries for As-Cast 13% Mn Steel Slabs (14)

The higher carbon grades show intergranular carbides and eutectic cells, in rough-cast heavy sections. This phenomenon is amplified by the presence of chromium. It is often difficult to identify clearly whether the intergranular carbides form in preference at the end of the solidification process, or whether they precipitate in the grain boundaries during cooling. Silicon, known for its embrittling effect in heavy sections, at contents of the order of 0.5 to 1.0% (15) probably plays a role by increasing the activity of carbon in the austenite. as well as influencing the kinetics of precipitation of intergranular carbides.

Phosphorus forms a low melting point eutectic ($\gamma + M_3P + M_3C$) (964°C) which appears in the primary grain boundaries at a variable phosphorus content according to grain size. This can be seen in thin

Name: Norm	Name: Commercial	Chemical Composition: C%	Mn%	Si%	P%	S%	Cr%	Mo%	V%	Ni%	Al%	N%
a - Abrasion Resistance												
ASTM A128A		1.05-1.35	11.0									
ASTM A128B1		0.9-1.05	11.5-14.0									
ASTM A128B2		1.05-1.20	11.5-14.0									
ASTM A128B3		1.12-1.28	11.5-14.0									
ASTM A128B4		1.20-1.35	11.5-14.0									
ASTM A128C		1.05-1.35	11.5-14.0				1.5-2.5					
ASTM A128D		0.70-1.30	11.5-14.0							3.0-4.0		
ASTM A128E1		0.70-1.30	11.5-14.0					0.9-1.20				
ASTM A128E2		1.05-1.45	11.5-14.0					1.8-2.10				
ASTM A128F		1.05-1.35	6.0-8.0					0.9-1.20				
ASTM A128G		1.00-1.50	15.0-18.0				3.0					
b - Railway Applications												
UIC		0.95-1.30	11.5-14.0				Mn >10 C					
c - Non-Magnetic Applications												
ASTM A289B		0.50	18.0	0.42	8	2.5	5.0			≤ 2.0		
DIN-X40MnCr18		0.40	18.0	0.55			4.0					
DIN-X55MnCr(N)18K		0.55	18.0	0.50	50	10	4.5		0.12	1.0		≥ 0.05
	RM711	0.70	15.0	0.28	45	04	0.24			1.1		
	RM720	0.71	15.7	0.32	47	03	0.21		0.59	1.2		
	RM711FW	0.72	15.6	0.25	15	51	0.25			1.1	0.011	Ca‰=2.5
	Non-magn 30	0.60	14.0	0.30			2.0			2.0		
	AM2	0.46	19.2	0.34	20	01	5.3					
	0.9C 14Mn	0.90	14.2	0.43	50	05						
	18M1	0.50	18.5	0.43	14	06	2.48			1.22		
	NM.11	0.74	15.9	0.40	24	05	1.82					
	MSE	0.60	17.0	1.0			6.0		2.0			
d - Cryogenic Applications												
	RM801FW	0.60	24.6	0.21	48	8	6.0					
e - Corrosion, Oxidation and Abrasion Resistance												
SAE EV11	SIL 746	0.7	6.3	0.56			21.0			1.9		0.23
SAE EV12	21.2 N	0.55	9.0	0.25			21.0					0.30
SAE EV13	Garman H	0.53	11.5	2.6			21.0					0.40
	21.4 N	0.50	9.0	0.25			21.5			4.0		0.40
	G 192	0.60	8.5	0.50			22.0	3.0				0.35
GOST20G28Yu9MBV		0.90	28.0								9.0	
		0.8-1.0	28.0-32.0	0.8-1.0							8.0-11.0	
GOST45G17Yu3		0.45	17.0				14.0-16.0	0.30			3.0	
		≤ 1.0	5.0-10.0									

Table 3 : Chemical Composition of Commercial Maganese Austenistic Steels

	Chemical Composition % (wt)				
	C	Mn	Fe	Mo	P
Steel	1.32	13.08		2.05	0.062
Oval Eutectic	1.13	24.20	60.60	5.70	7.69
Lamellar Eutectic	1.35	25.20	57.20	7.54	5.31
1 hr at 1100°C	1.38	18.32	61.15	16.79	1.48
(Si, S and Cr have not been analysed)					

Table 4 : Chemical Composition of Phosphide Eutectics (16)

	Metallographic Evidence of Liquation				
Sample Composition, % and Section Size	**1000°C**	**1030°C**	**1050°C**	**1100°C**	**1150°C**
1.2C-12Mn-.064P; 16 mm section, lab heat	None	Isolated Regions	Isolated Regions	Isolated Regions	Several Grain Boundary Regions
1.34C-12Mn-.055P; 16 mm section, lab heat	None	None	Isolated Regions	--	--
	1095°C	**1150°C**	**1205°C**	**1260°C**	
1.0C-7Mn-1Mo-.03P; 75 mm section; lab heat	None	None	None	Isolated Regions	
1.0C-7Mn-1Mo-.08P; 75 mm section, lab heat	None	None	Isolated Regions	Several Grain Boundary Regions	
1.1C-7Mn-1Mo-.03P; 175 mm section; lab heat	None	None	None	Isolated Regions	
1.0C-7Mn-1Mo-.09P; 175 mm section, lab heat	None	None	Isolated Regions	Several Grain Boundary Regions	
1.3C-12Mn-0Mo-.04P; 150 mm section; commercial	None	None	None	Extensive	
1.4C-12Mn-1Mo-.04P; 75 mm section, lab heat	None	Several Grain Boundary Regions	Several Grain Boundary Regions	Extensive	
1.2C-12Mn-2Mo; 250 mm section, commercial	None	None	Several Grain Boundary Regions	Extensive	

Table 5 : Response of Austenistic Manganese Steels to Heat Treatment (15)(17) section castings at a content of 0.12% phosphorus and at less than 0.07% phosphorus in heavy section castings. The phosphorus eutectic has different morphologies according to phosphorus and molybdenum contents: with low phosphorus contents, it has an oval shape and the interior of the eutectic structure is surrounded by a carbide layer from which carbide needles extend. This layer and the carbide needles disappear in the presence of molybdenum, and the eutectic takes on a lamellar form a few microns in width.

The chemical composition of the oval and lamellar eutectic cells differs, the oval cells being richer in phosphorus, and the lamellar cells being richer in carbon and molybdenum (table 4). The needles and the external layer of carbide disappear with the increase in phosphorus content, and the phosphorous eutectic forms a continuous network as soon as the phosphorus content is sufficiently high (about 0.29% P in thin section casting).

The segregation of phosphorus, carbon and molybdenum can cause liquation phenomena, during heat treatment, in steels where these elements are found in relatively high nominal contents (table 5). The higher carbon grades containing chromium, which display eutectic cells in the as-cast state, are also susceptible to liquation phenomena due to higher solution treatment temperatures.

c) Austenite transformation in isothermal and continuous cooling conditions

Transformation in isothermal conditions

The transformation of manganese austenites in isothermal conditions has been studied only in Hadfield and related steels, and in metastable austenites with 6% Mn. The austenites under consideration are supersaturated in carbon and undergo the transformation $\gamma \rightarrow \gamma$ + carbides below Acm, that is, between about 300°C and the Acm temperature. Decomposition of the austenite initially produces an intergranular carbide which is the first to form, and tends towards a continuous network; an acicular carbide which precipitates in the grain in the form of a carbide-ferrite aggregate (at high temperature) and, solely in the carbide state at low temperature, pearlite; finally, metastable austenites after cooling show martensite, next to carbides and the pearlite.

The transformation of Hadfield steel in isothermal conditions, between 300 and 700°C (figure 8a) begins with the precipitation of intergranular carbides after a few seconds of incubation time. Next, the acicular carbides appear after an incubation period described by a C-like curve, of which the nose is located at a temperature and an incubation time of the order of 600°C and 1 min. respectively. The pearlite appears after a longer incubation time, and in a narrower temperature range than that of the acicular carbides. The pearlitic transformation is very slow: at the nose of the curve where the kinetics are the most rapid, after holding at a temperature for a very long time, not more than 30-50% pearlite is obtained.

Chemical composition has a strong effect on the transformation kinetics of manganese austenites. The incubation time of intergranular carbide precipitation at 600°C in Hadfield steel, jumps from a few seconds to 2 minutes when carbon content decreases from 1.27 to 0.94% (figure 8b). The incubation time curve for intergranular carbides towards short times is of asymptotic shape; for a carbon content of 1.3%; it takes on a C shape at low levels of carbon, which is in agreement with the results of Saito (21) who found that the incubation time for the precipitation of intergranular carbides of an austenite at 0.9% C -1% Si - 14.36% Mn, is around 2 min. at 600°C and 5 min. at 800°C (figure 9).

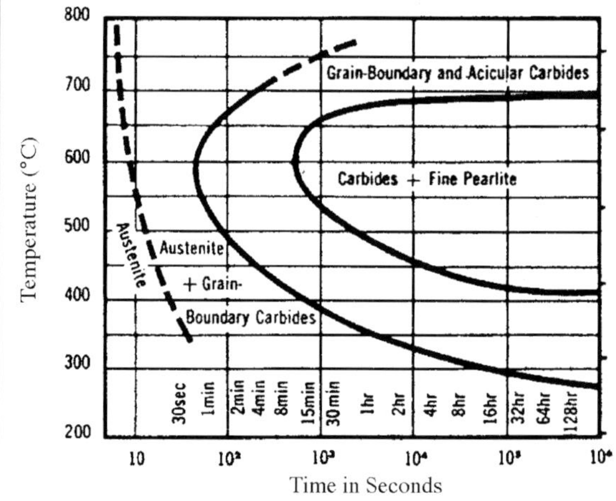

Figure 8a : Isothermal Transformation Diagram of a Steel with 1.28%C. 12.4%Mn 0.35%Si 0.009%S. 0.30%P. 0.28%Ni and 0.23%Cu. Austenitized 30 min. at 1050°C. Grain Size 5 to 6 (18)

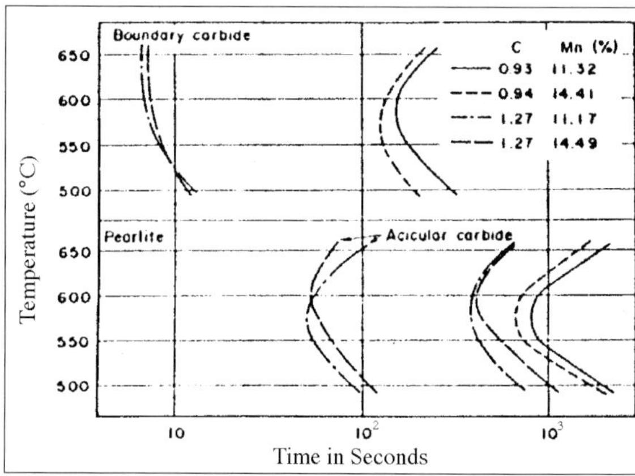

Figure 8b : Effect of C and Mn on the Begining of Isothermal Transformation of 13% Mn Cast Steels (19)

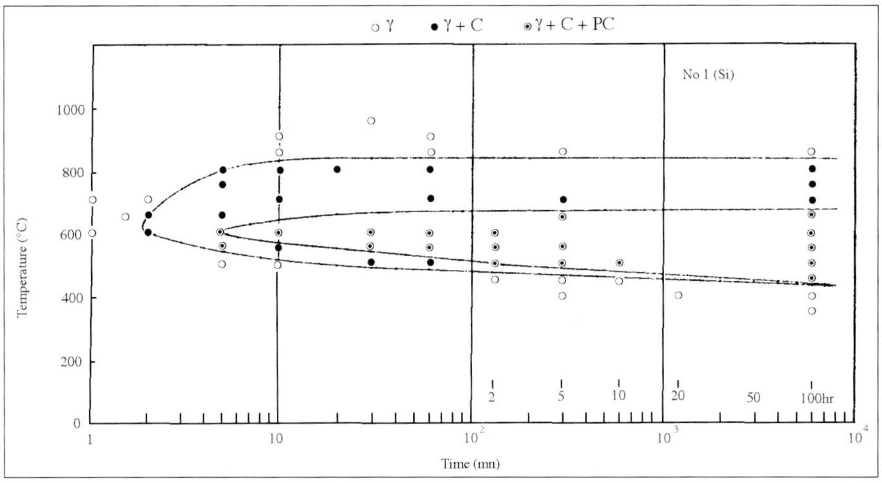

Figure 9 : Isothermal Transformation Diagram after Solution Treatment (30 min at 1100°C) (0.95%C - 1.08%Si - 14.36%Mn) (21)

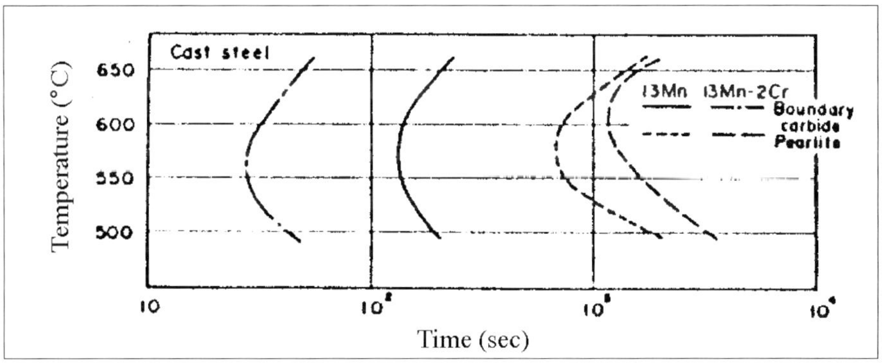

Figure 10 : Effect of Cr addition on the Begining of Isothermal Transformation of 0.95%C - 13% Mn Steels (19)

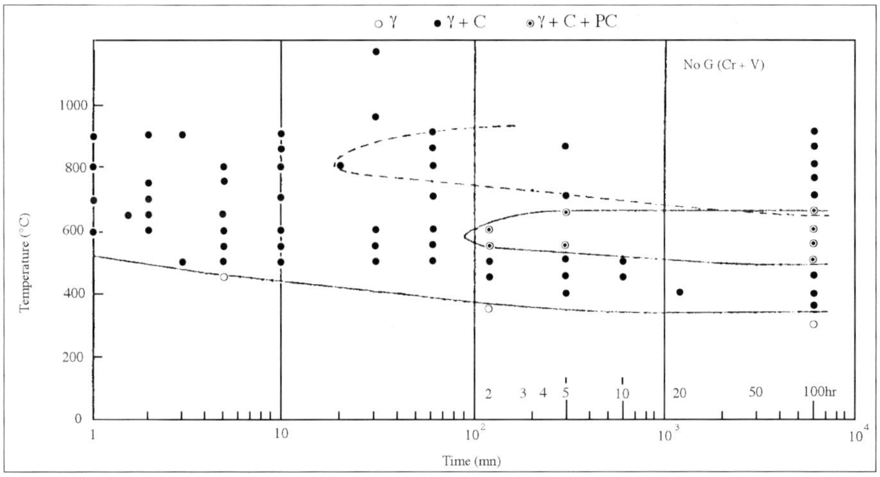

Figure 11 : Isothermal Transformation Diagram after Solution Treatment (0.86%C - 1.05%Si - 10.03%Mn - 4.06%Cr - 0.55%V) (30 min at 1200°C)(12)

Variations in manganese content of the Hadfield steel, of the order of 3%, have little effect on the incubation time of the intergranular carbide precipitation (figure 8b). At a manganese content ofthe order of 6% and carbon content of between 1.1 and 1.36%, very few intergranular carbides can be seen. After isothermal treatment, most are fine, isolated carbides with no tendency to form a continuous film. The inhibiting effect of a lower manganese content on the precipitation of intergranular carbides, is probably connected with a lower carbon oversaturation of the austenite, the eutectoid composition being 0.56% C at 6% Mn and 0.39% C at 12% Mn. Chromium reduces the incubation time of the intergranular carbide precipitation (figure 10), and vanadium would have the same effect as chromium (figure 11). On the other hand molybdenum increases incubation time (figure 12) as does aluminium.

An increase in carbon content in austenites at 6% Mn, shortens the incubation time for acicular carbide precipitation, displaces the temperature of the nose of the transformation curve to a higher one, and widens the domain of the carbide precipitation. The carbide precipitation is rapid when it starts, quickly attaining an asymptotic value when reaching 10 to 14% carbides by volume. Molybdenum inhibits the formation of acicular carbides, its effect being comparable with a decrease in carbon content (figure 13).

The incubation time of the pearlitic transformation of Hadfield steels is reduced by an increase in carbon content, and extended by an addition of molybdenum, chromium and vanadium (figures 10-12). Steels at 6% Mn can be distinguished from Hadfield steels with respect to pearlitic transformation, by a reduced effect of the carbon content on the incubation time, and mainly by more rapid transformation to pearlite. Complete transformation of the austenite occurs within about 5 hours, at 550-600°C. Molybdenum extends the incubation time and slows the pearlite transformation kinetics (figure 13).

Austenite at 30-33% Mn - 8-10% Al - 0.03-1.0% Si - 0.9-1.3% C undergoes a transformation between 500 and 700°C. In a primary phase, a modulated structure with areas poor and rich in carbon, is formed by spinodal decomposition. The consistency of the

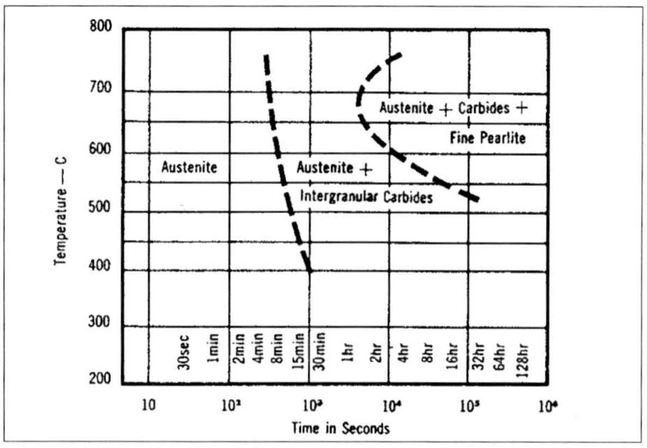

Figure 12a : Influence of a 2% Molybdenum Addition on the Isothermal Transformation of an Austenitic Manganese Steel (19)

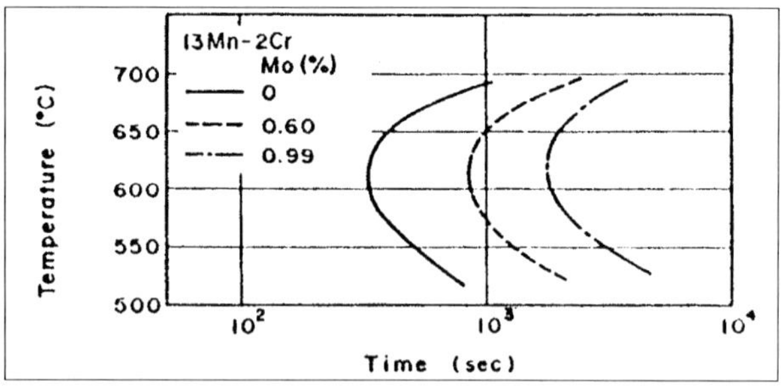

Figure 12b : Effect of Mo on the Beginning of the Isothermal Pearlite Transformation in 13%Mn - 2%Cr Cast Steel (19)

deformation caused by a modulation of the microstructure produces a significant hardening. The carbide $K(FeMn)_3AlC_x$, (structure $L'1_2$) then precipitates in carbon-rich areas and finally, after maintaining the temperature a long time, βMn phase and ferrite form in the grain boundaries (22) (23).

Transformation during continuous cooling

The transformation of Hadfield steel during continuous cooling after solution treatment is localized near the grain boundaries. It consists of intergranular carbide precipitation and transformation in the pearlitic zone followed by a bainitic transformation. No acicular carbide precipitation is observed in grains as is the case during isothermal treatment.

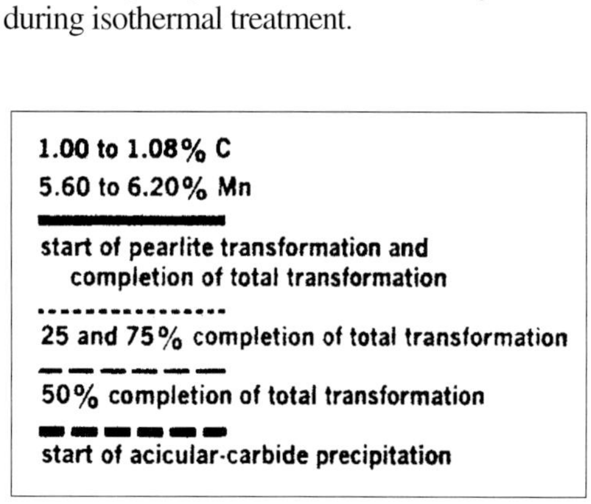

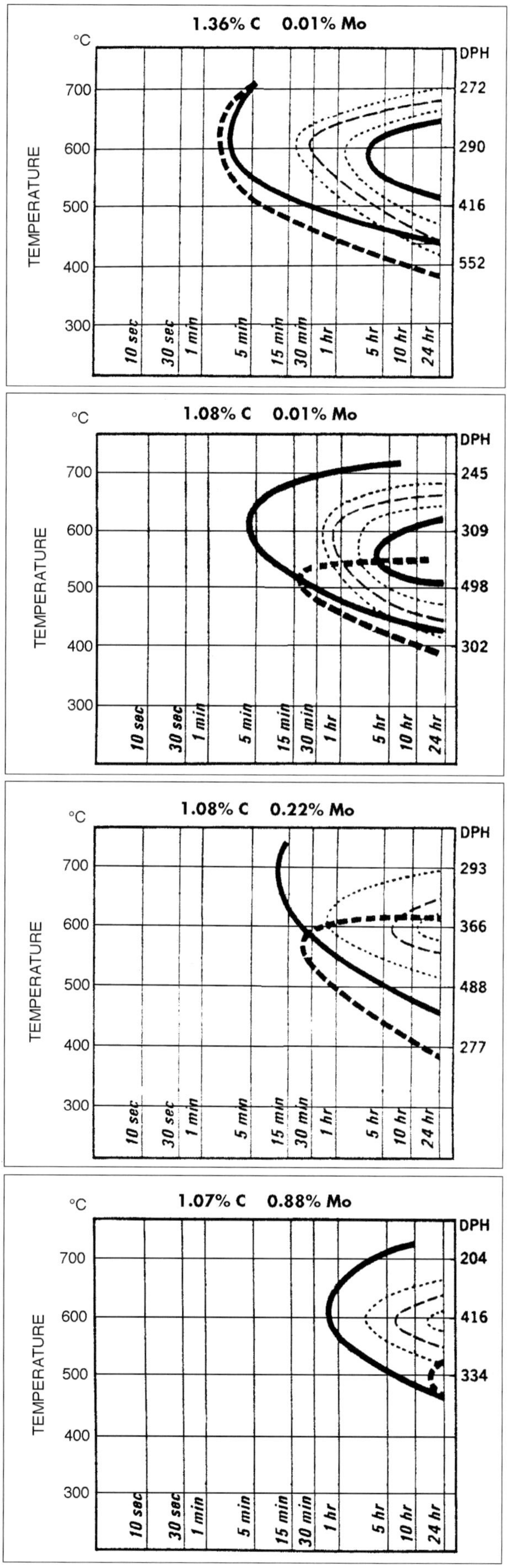

Figure 13 : Effect of C and Mo Content on the transformation Kinetics under Isothermal Conditions, of 5,6 to 6.2%Mn Austenites (20)

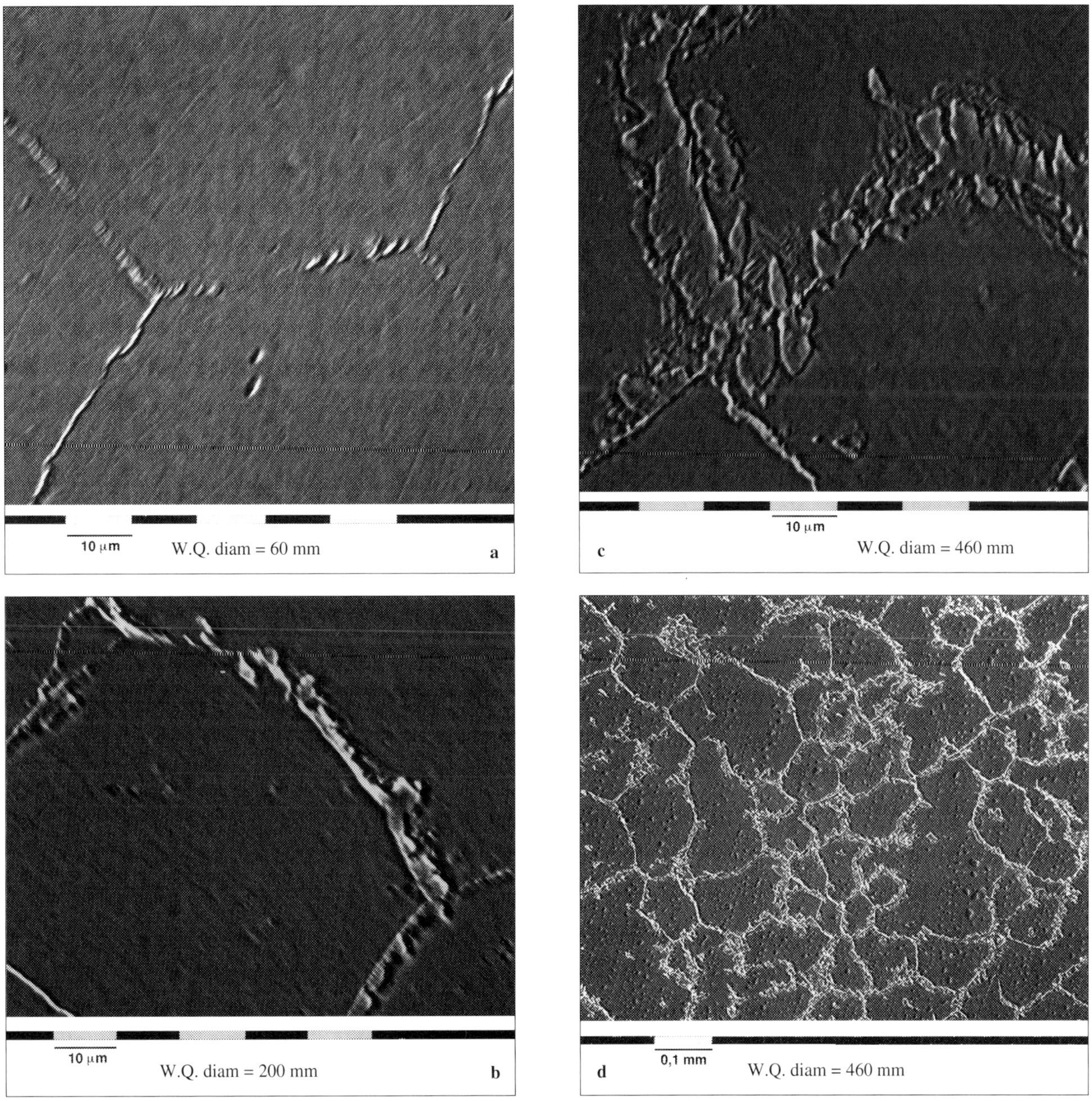

Figure 14 : Influence of the Section Size on the Microstructure of Rolled Hadfield Steel (1.10% C - 0.25% Si - 12% Mn)
Solution Treatment 40 min at 1100°C Water-Quenched

Before the intergranular carbide precipitation there is a widening of the grain boundaries, which probably corresponds to a localized concentration of carbon, and is a preliminary step to carbide precipitation (figure 14a). The carbides then precipitate in the grain boundaries in the form of platelets about 0.1 μm thick. The density of precipitation increases progressively when the cooling rate is lower, and the carbides finally form a continuous network, 0.2-0.5 μm thick (figure 14b).

A transformation occurs in the pearlitic area, at relatively slow cooling rates, and is followed, at a lower temperature, by a bainitic transformation localized in a narrow area poor in carbon, at the edge of transformation products (figure 14c). The microstructure of the cells formed in the pearlitic area has not been resolved with an electron microscope : it has unidentified and rather indistinct uniform lamellar areas (figure 15). The carbon content in these cells is similar to that of the austenite matrix (figure 16) and magnetic measurements show the presence of a large volume fraction of ferrite which leads to the assumption that the transformation product would be an aggregate of ferrite and very fine carbides.

The addition of about 1% molybdenum to a Hadfield steel does not alter the morphology of the transformation products on continuous cooling ; the addition of 1.4% chromium shows a tendency towards the formation of lamellar cells (figure 17). The combined effect of higher manganese and chromium contents (27.6% Mn, 2.24% Cr) and of a sufficiently high percentage of carbon produces, after continuous cooling, two lamellar structures of

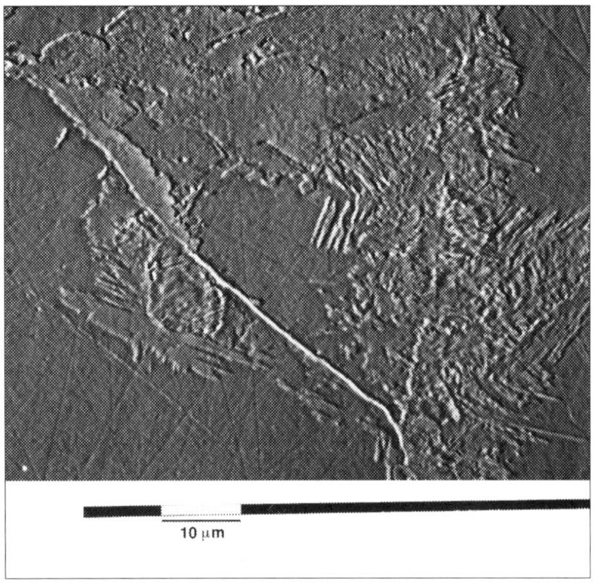

Figure 15 : Microstructure Obtenaid by Continuous Cooling after Traversing Pearlitic Domain
(1.14% C - 0.38% Si - 13.7% Mn - 0.91% Al)
(40 min 1100°C -Water-Quenched 640 mm diameter)

pearlitic appearance (figure 18). One of these forms between 650 and 375°C, the other forms at lower temperatures, between 400-200°C. The two structures differ from pearlite by the absence of ferrite, and the aggregates are not ferromagnetic; the manganese content of the matrix remaining sufficiently high to allow the interlamellar spaces to remain austenitic with a low carbon content. The two aggregates very likely differ one from the other by the type of

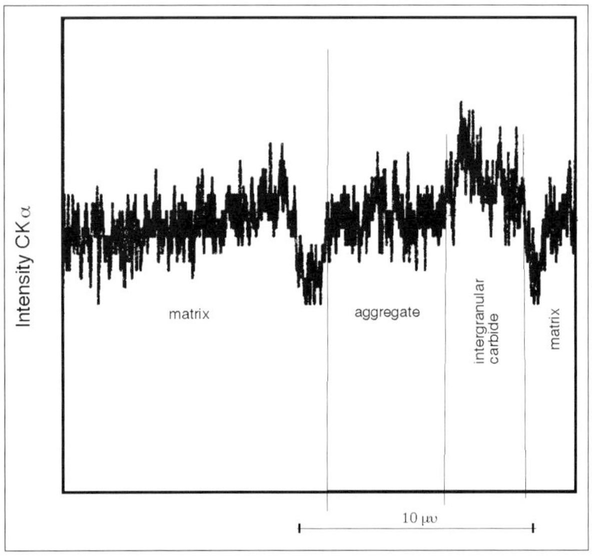

Figure 16 : Carbon Distribution in the Matrix and Transformation Products of Austenite at 1.11% C - 0.25% Si - 12.0% Mn
(40 min at 1100°C -Water-Quenched 640 mm diameter)

carbide, which for the structure present in the pearlitic area would be the carbide M_3C, and the carbide M_7C_3 or $M_{23}C_6$ for the structure formed at a lower temperature.

A steel of composition 1.27% C - 13% Mn - 0.97% Mo can retain, at the same critical diameter of 200 mm, a higher supersaturation rate of carbon than the steels at 1.0-1.10% carbon.The critical section for the

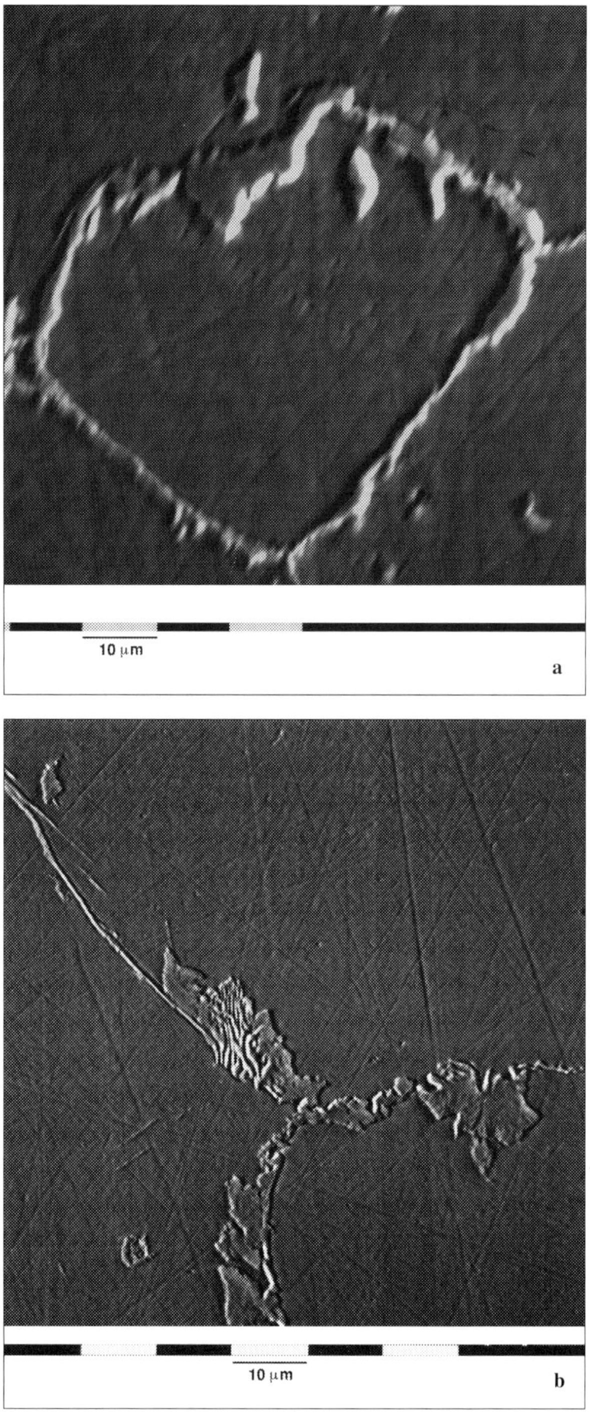

Figure 17 : Microstructure of Manganese Austenitic Steels with Addition of Cr and Mo Obtained after Simulation of Cooling Cycle in a 770 mm Diameter Cylinder Air-Cooled
(Solution Treatment 40 min at 1100°C)
a) 1.27% C - 0.45% Si - 12.9% Mn - 0.965% Mn
b) 1.18% C - 13.6% Mn - 1.4% Cr

appearance of a continuous carbide network is considerably higher than that of steels without molybdenum and a lower carbon content, being 640 mm as against 340 mm. The transformation in the pearlitic area is displaced about 100°C towards lower temperatures and the critical diameter increases from about 450 to 750 mm. A bainitic transformation has not been detected, probably due to the slow transformation rate (figures 19-20).

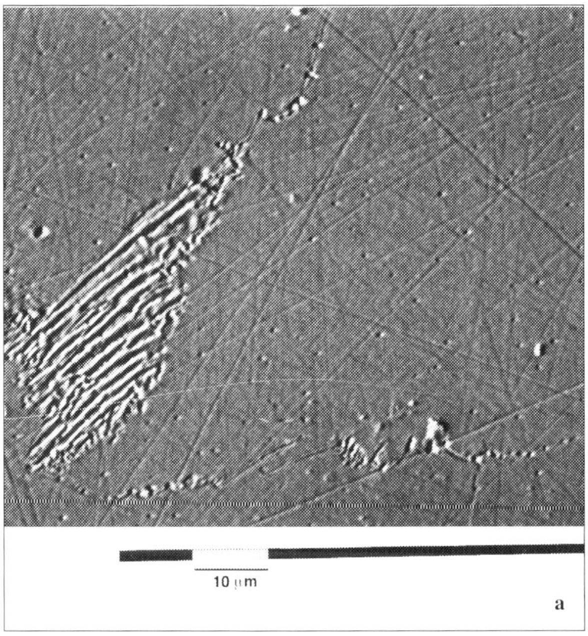

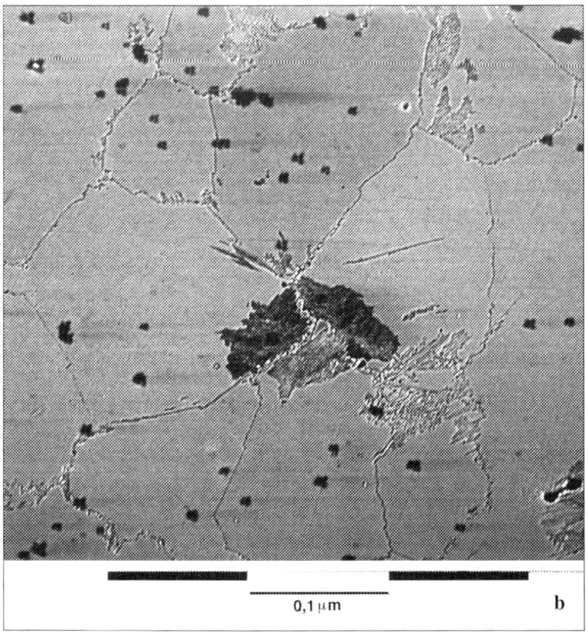

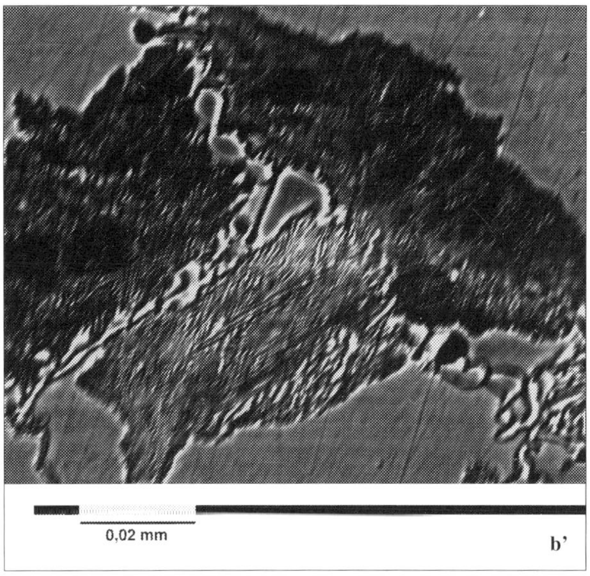

Figure 18 : Lamellar Microstructure of Steel with 1.34% C - 0.8% Si - 17.6% Mn - 2.2% Cr obtained by Continuous Cooling Though the Pearlitic Area (solution Treatment 40 min at 1100°C)
a) $M_3C + \gamma$ - water-quenched 320 mm diameter
b) $M_3C + \gamma$ and $\gamma + M_7C_3$ or $\gamma + M_{23}C_6$ - water-quenched 640 mm diameter

A decrease in manganese content to 6.6% of a steel containing 1.27% C and 0.97% Mo does not alter the critical section for the precipitation of the intergranular carbides. Transformation in the pearlitic area is raised again by 100°C, and the critical sections of the continuous carbide and "pearlitic" network are lowered to diameters not much greater (220 mm) than the critical precipitation section of the intergranular carbides (200 mm). The bainitic transformation, which succeeds the transformation in the pearlitic area, occurs over a relatively wide range of temperature, and is followed by a martensitic transformation.

A steel with the composition 1.18%C- 13.6%Mn-1.4%Cr has the same critical section for the precipitation of intergranular carbides as steels with 1.0-1.10% carbon, and a slightly higher critical section for the continuous carbide network formation. The area of pearlitic transformation, similar to that of the molybdenum steel, is displaced by 100°C to lower temperatures, and the critical "pearlitic" diameter is clearly greater than that of the steel at 1.0-1.10% C, but does not equal that of the molybdenum steel.
The continuous cooling transformation (CCT) diagram of steel D at 1.34% C - 17.6% Mn - 2.2% Cr (figure 20) is a different shape from the CCT diagram of steel C which has lower carbon, manganese and chromium contents. The critical sections for the precipitation of the intergranular carbides of the continuous carbide and pearlitic network of this steel, are all close to 180 mm. The starting "pearlitic" transformation temperature decreases with increase of the section, following carbon depletion of the austenite due to an increase in the precipitation of intergranular carbides during cooling, and to the strong effect of carbon on the kinetics of the "pearlitic" transformation.

Examination of the microstructure obtained after cooling which simulated the thermal cycle of air-quenching cylinders of different sections, showed (figure 21) that steel with 1.6% C - 12.6% Mn - 3% Al retained the carbon in solution in the austenite up to a section of about 250 mm, after a solution treatment of 30 mm at 1150°C, followed by air-quenching. A structure of pearlitic appearance, containing no intergranular carbides and unresolvable by optical microscopy, appears at lower cooling rates. The addition of aluminium effectively inhibits the precipitation of intergranular carbides in the steel under consideration, unlike a steel with 1.14% C - 12.9% Mn - 0.9% Al (figure 19c), where the aluminium addition is probably insufficient taking into account the carbon content.

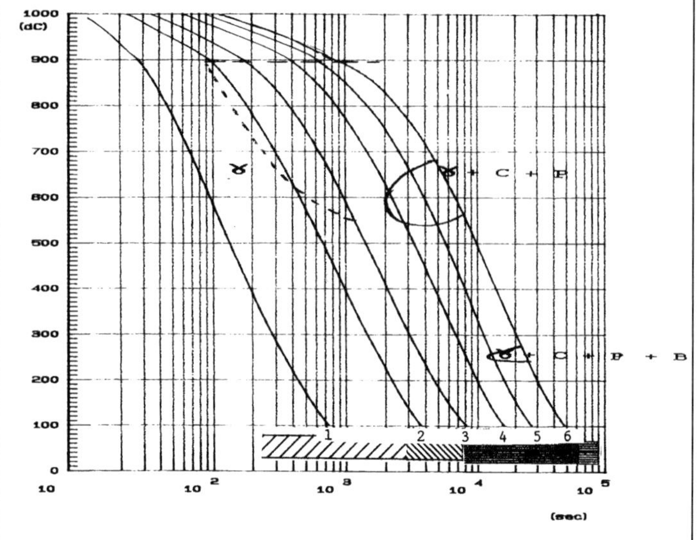

a

b

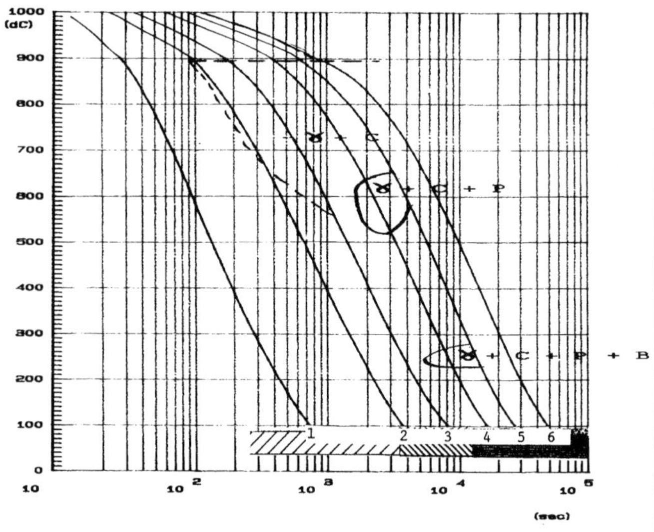

c

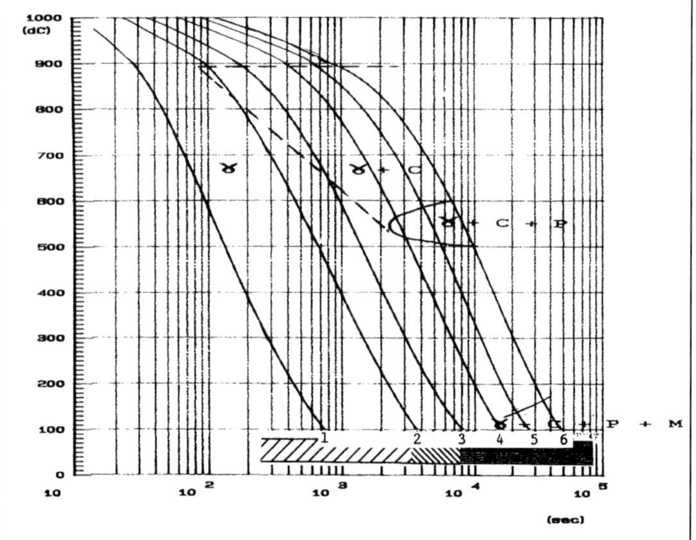

Figure 19 and 20 :

1) Thickness of flat product equivalent to critical diameter :
Dia 180 mm → e = 125 mm
Dia 320 mm → e = 210 mm
Dia 455 mm → e = 295 mm

2) Cooling curve - W.Q. diameter (mm)

1	90
2	200
3	320
4	460
5	640
6	770

Figure 19 : CCT Curves for Hadfield Steels
(Solution Treatment 40 min 1100°C)

	C%	Si%	Mn%	
a)	1.10	0.25	12.0	Rolled Plate Steel
b)	1.02	0.38	13.37	Cast Steel
c)	1.14	0.37	12.94	(+Al=0.91%) Rolled Steel

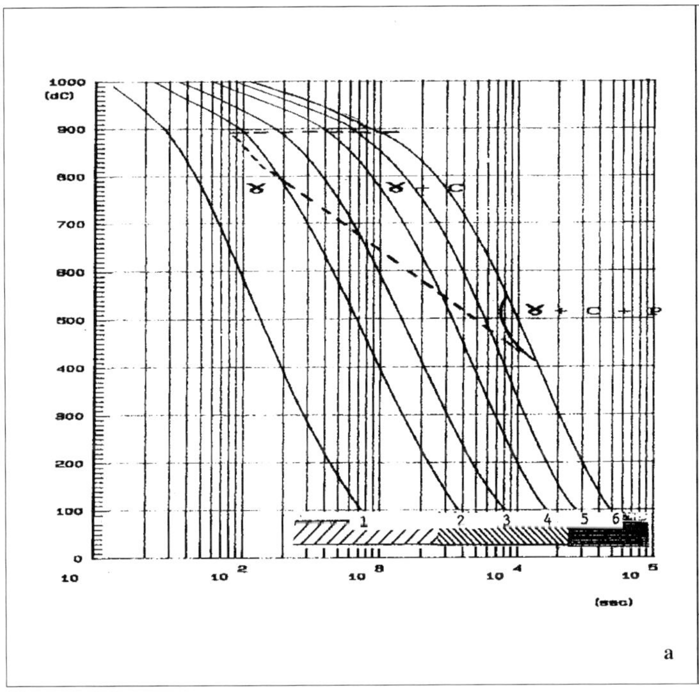

a. 1.27% C - 0.45% Si - 12.9% Mn - 0.965% Mo

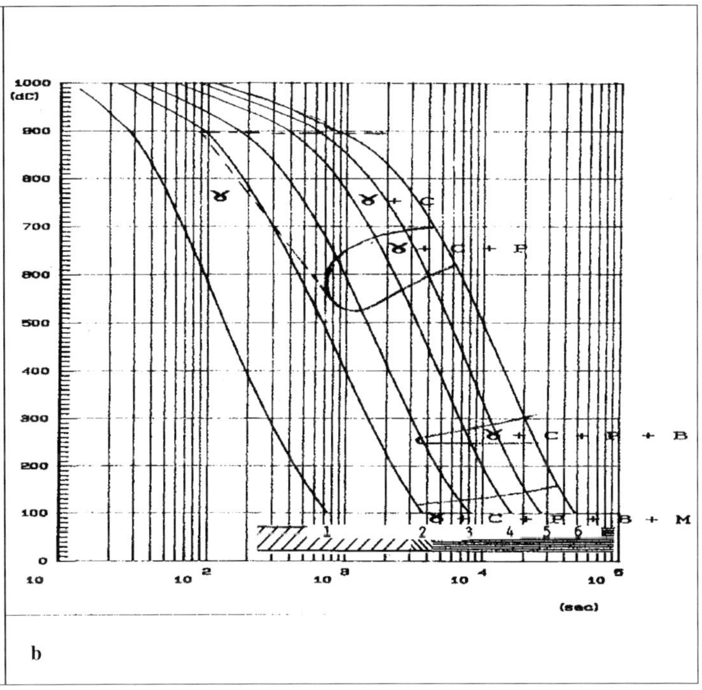

b. 1.24% C - 0.42% Si - 6.64% Mn - 0.98% Mo

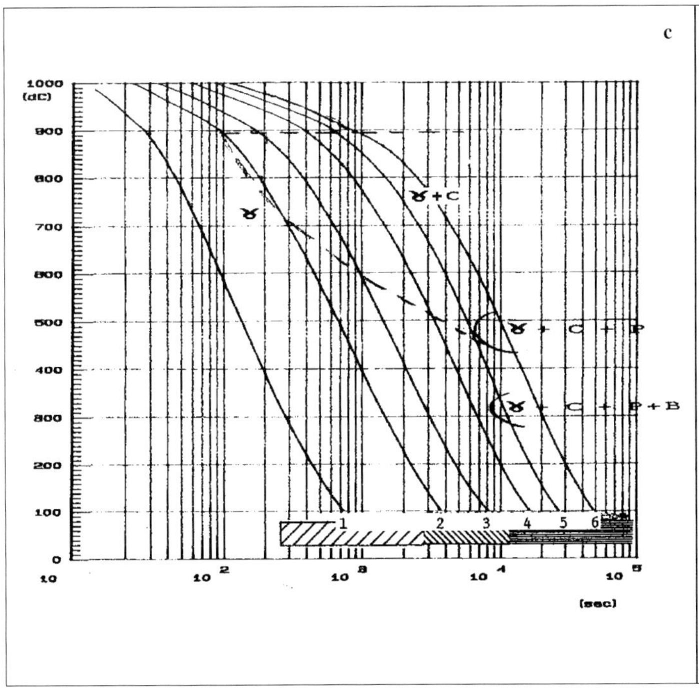

c. 1.18% C - 0.39% Si - 13.65% Mn - 1.42% Cr -

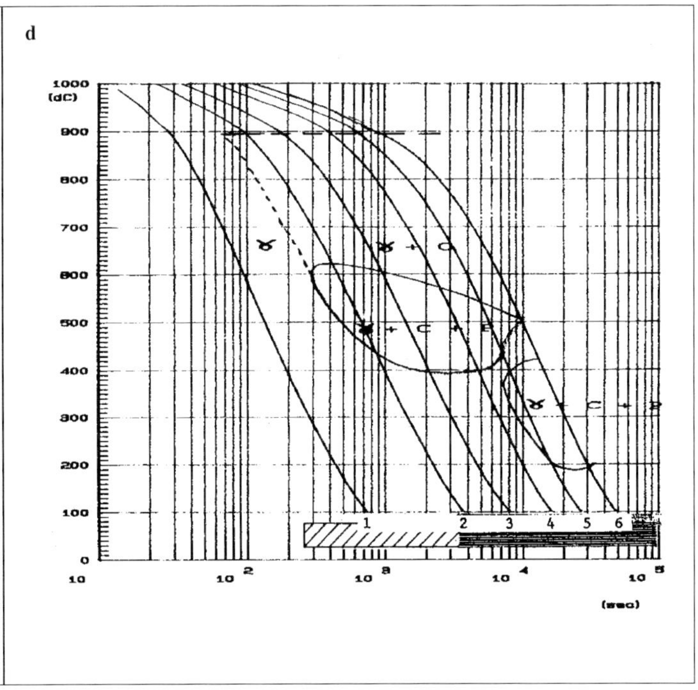

d. 1.34% C - 0.8% Si - 17.6% Mn - 2.2% Cr

Figure 20 : Continuous Cooling Transformation (CCT) Diagram of Manganese Austenitic Steels
- Effect of Mn, Cr and Mo (Cast Steels) (Solution Treatment 40 min 1100°C)

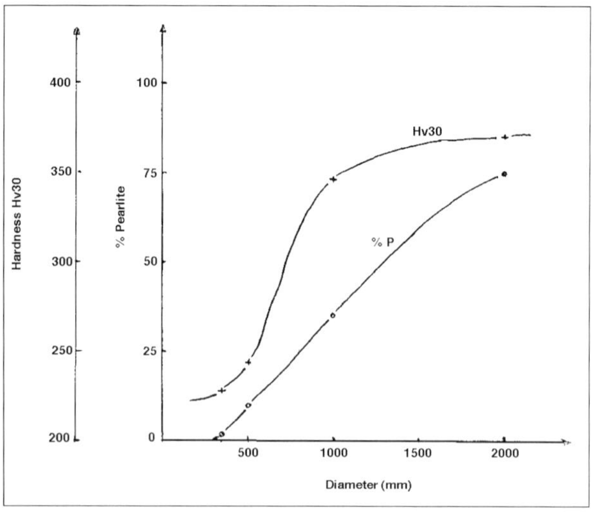

Figure 21 : Influence of Quenched Section Diameter on Microstructure and Hardness of Steel with 1.6% C - 12.6% Mn - 0.51% Si - 3.0% Al (30 min 1150°C Air-Cooled)

Ferrous-base alloys with high chromium and carbon contents have a strong propensity to retain austenite at high cooling rates: the addition of manganese provides an austenitic structure in the quenched state, in heavy sections. Figure 22 gives values of the C-Mn pair, which produces, in an alloy at 12% chromium, an austenitic structure in sections of 350 to 2000 mm in diameter after air-quenching.

Most of the transformation curves were established using test pieces taken from small section ingots, or from rolled products made at a high rate of rolling. They are representative of the response of a relatively homogenous austenite to heat treatment, and should be used with caution. According to the CCT diagram

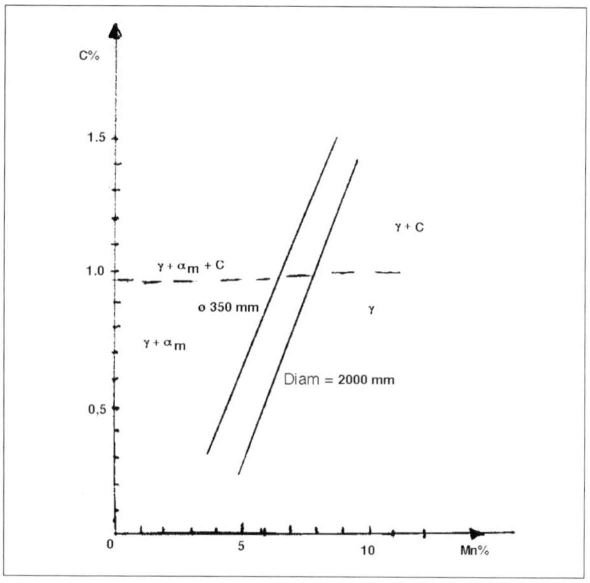

Figure 22 : Effect of Carbon and Manganese Contents on the Limit of Domains $\gamma / \gamma + \alpha_m$ of an Alloy Fe - 14% Cr for Critical Sections of 350 and 2000 mm. (24)

in figure 20c, the addition of 1.4% chromium leads to a large critical section for quenching, which is in contradiction to industrial practice. Transformation of the austenite is localized near the grain boundaries. The transformation kinetics do not depend on the nominal chemical composition of the alloy, but rather on that of the interdendritic spaces which result from casting segregation. In reality, a steel of the same chemical composition cast into thick sections has, because of segregation, a lower critical quenched section which approximates to that of steel with higher carbon and chromium contents as seen in figure 20d.

II. Work-hardening Mechanisms

a) Deformation process, crystal defects

The yield stress of manganese austenites does not show any singular point on the stress-strain curve and, if the alloy's stacking fault energy is sufficiently low, stacking faults appear in the grains appropriately oriented in the direction of the stress, as soon as the elastic limit is passed. The stacking faults thicken and transform into twins at a low deformation rate (figure 23). Twins appear in a second twinning system at high deformations, and twins have been observed meeting at right-angles, blocking each other (26). The grains thus divide up into small areas and their boundaries form obstacles to dislocation movement, as does grain refining. The grains near the fracture present shearing bands which intercept the twins at a 36° angle.

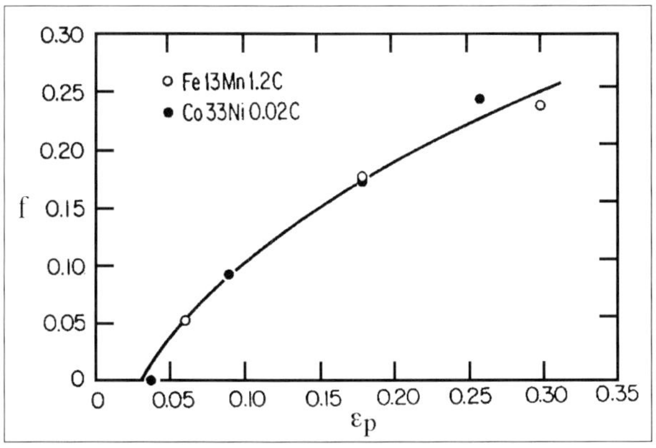

Figure 23 : Twinned Volume Fraction. *f, vs* True Plastic Strain for Hadfield Manganese Steel (Open Circles) and an fcc Co-33% Ni-0.02% C Alloy. Both Alloys were Deformed in Tension at Room Temperature. Hadfield composition is 1.18% C - 12.9% Mn - 0.2% Si (25)

A high dislocation density at the interface twin-matrix as well as dislocation clusters have been observed using transmission electron metallography (26) (27). The X-ray lines become larger when the deformation increases, as the number of crystalline defects increases (28).
Twinning deformation is reduced or eliminated following the stacking fault energy of austenite and the deformation temperature, therefore deformation proceeds more or less by slipping. Slips occur frequently on two slip systems producing intersecting slip lines. The slip lines do not present any cross slips, and are not in uniformly spaced bands as in aluminium and are thicker near the grain boundaries than in the grain(30). The slip-bands and the twins are 0.005-0.3 μm and 0.005-10 μm thick respectively, and are often difficult to distinguish from each other under the optical microscope, especially when they are thin. Deformation is heterogenous at both the macroscopic and microscopic level, until fracture. The deformation of the specimen is the average of the deformation of the grains, but the deformation of individual grains can be more or less important (29).

b)Redistribution of Carbon Atoms

Several methods of investigation such as tensile tests, Mössbauer spectroscopy and X-ray diffraction, revealed a redistribution of carbon atoms at more or less short distances apart, during deformation.
The stress-strain curve of Hadfield steel and metastable austenites, which have an Ms point below room temperature, present a serrated flow caused by dynamic strain-aging during deformation (figure 24). This phenomenon can be explained by the reorientation of the Mn-C pairs, or of the C-Mn-C combinations, in the dislocation stress field, caused by the short-distance "jumping" of carbon atoms which pin together the dislocations and lower their density. As the deformation rate is imposed, and remains constant, the applied stress increases in order to restore the dislocation density ; the new dislocations are displaced in the network, reduce their displacement rate, and the applied stress diminishes. The dislocations are rapidly pinned together, and the process is repeated.

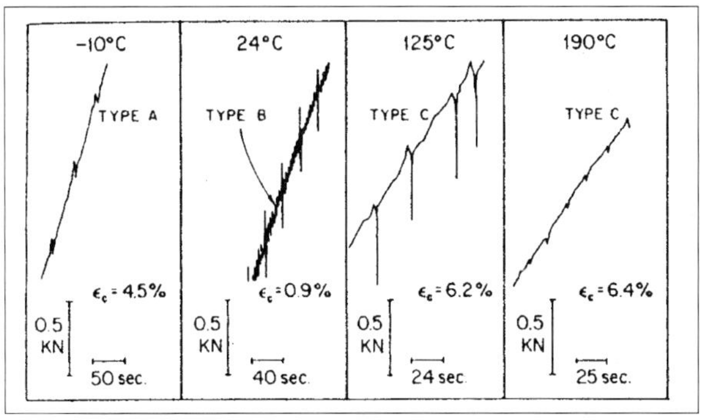

Figure 24 : Serrated Flow in Hadfield Steel as Affected by Temperature Nominal Strain Rate 3 x 10^{-4} s^{-1} ε_c = "critical Strain" (31)

The apparent activation energy of the serrations (104 ± 15KJ mol^{-1}) (figure 25) is considerably less than the diffusion energy of the carbon in Hadfield steel (140 - 150 KJ mol^{-1}), and it corresponds to the activation energy of diffusion of the carbon, over short distances, needed to pin a dislocation, and is 0.6 to 0.7 times the carbon bulk diffusion.

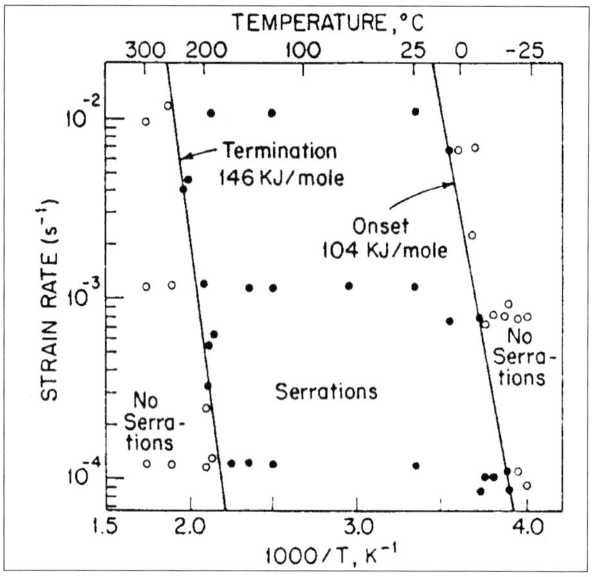

Figure 25 : Strain Rate-Temperature Regime of Serrations in Hadfield Steel (31)

Mössbauer spectroscopy has clearly indicated the formation of clusters of carbon atoms (32) and an increase of 50 to 60°C in the Néel temperature after work-hardening (figure 26) (28) (33).

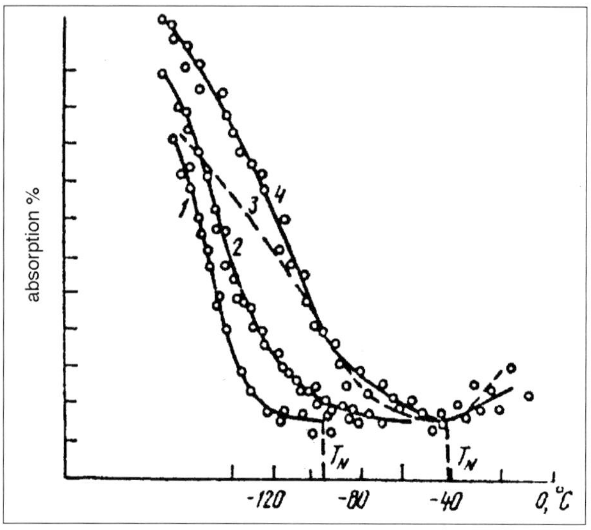

Figure 26 : Global Absorption, Mössbauer Spectra, at Different Temperatures, of the Steel - 1/10% C - 13% Mn (33)
1- Quenched at 1100°C 2- Rolled - 30% Reduction
3- Shearing Deformation (ε = 6.9) 4- Surface Machined

The Néel temperature characterizes the transition from the paramagnetic state to the non-ferromagnetic state : increasing this temperature could cause a displacement of the carbon atoms out of the iron-carbon pairs towards the manganese atoms, thereby increasing the magnetic field, which explains the change in the magnetic state.

The crystal lattice of the austenite in Hadfield steel scarcely changes until deformation of the order of 10 to 15% is applied, and decreases rapidly at higher deformation and tends towards a constant value from a deformation of the order of 40% onwards (figures 27 and 45) (28) (34). The shrinking of the crystal lattice is attributed to the rejection of carbon in solution in the austenite (figure 28). It can produce a localized segregation in carbon.

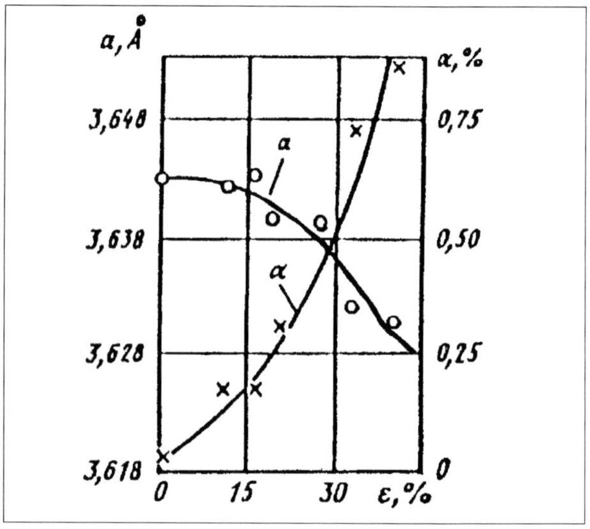

Figure 27 : Relationship between the Austenite Crystal Lattice, the Volume Fraction % of Phase α_m and the Deformation (Austenite at 1% C - 10% Mn - 0.2% Si) (34)

The residual carbon content of the austenite is sufficiently high to keep the microstructure of the Hadfield steel completely austenitic, as long as the dendritic segregation is not excessive.

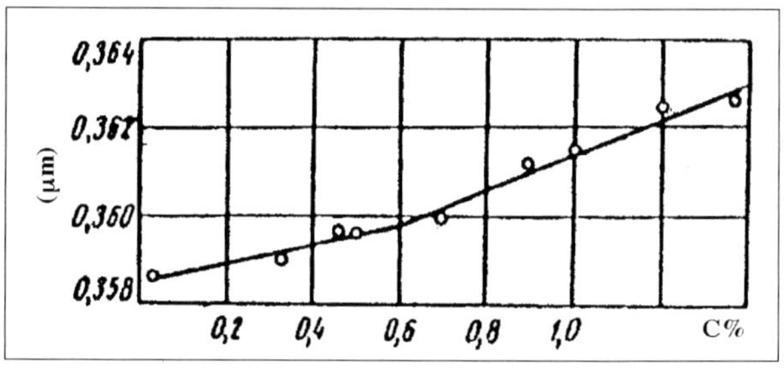

Figure 28 : Relationship between the Austenite Crystal Lattice and the C Content (12-14% Mn, Water-Quenched at 1200°C) (28)

It is difficult to build a model for carbon diffusion in a microstructure with a heterogenous chemical composition, the high value of the carbon displaced in the solid solution would lead to the assumption of diffusion over a greater distance than in the mechanism implied by dynamic strain-aging. In support of this hypothesis, it can be noted (figure 25) that the activation energy of the serration phenomenon is not constant, and moves towards the activation energy for bulk diffusion of carbon into the austenite at increasing temperatures. If the deformation conditions change from isothermal to non-isothermal the diffusion process can change from short-distance diffusion to bulk diffusion.

c) Martensitic Transformation, Carbide Precipitation

Comparing figures 5 and 29 indicates the field of metastable alloys. The limit of the $\gamma + \varepsilon/\gamma$ areas is displaced towards higher manganese and carbon contents by taking over the γ zone. The extreme point of the $\gamma + \varepsilon + \alpha_m/\gamma + \varepsilon$ field is also displaced towards higher manganese and carbon contents, merging with the quenching structures $\gamma + \varepsilon$, the chemical composition of which enables the successive transformations $\gamma \rightarrow \varepsilon \rightarrow \alpha_m$ to occur during work-hardening (30).

The metastable state depends not only on carbon and manganese contents, but also on alloying elements, deformation rates and temperature. The domains $\gamma/\gamma + \alpha_m$ and $\gamma + \alpha/\gamma + \alpha + \varepsilon$ seem the least well-defined as shown by a few examples :

ε martensite, from an austenite of 13% Mn, 0.01% C, with a $\gamma + \alpha + \varepsilon$ structure in the quenched state, is completely transformed into the α_m phase when deformation is sufficiently high.

the austenite of 13% Mn, 0.4% C, presents traces of the α phase in the quenched state, and 20% of martensite α_m after deformation of 10%, as against 42.5% martensite α_m and 15% martensite ε after 25% deformation.

austenites with less than 10% Mn, and more than 0.6% C (amongst which are found industrial grades of the 6Mn-1Mo type, with a stable domain of the austenite,) are in fact metastable austenites which undergo the $\gamma \rightarrow \alpha_m$ transformation beyond a certain critical deformation.

The temperature range between points Md and Ms is considerably greater for manganese austenites than for nickel-chromium austenites, which represents a greater instability (table 6).

Grade	Temperature (C°) M_s	M_d	M_d-M_s
17.5 Cr-8.3 Ni-0.111(C+N)	-196	70	266
17.8 Cr-7.4 Ni - 0.034 (C+N)	-40	90	130
4.79 Mn- 0.29 Si - 1.1 C	-19	450	469
5.71 Mn - 0.31 Si - 1.03 C	-73	400	473
24 Mn - 0.26 C	-180	260	440

Table 6 : Temperature of M_S and M_d Points of Chromium-Nickel and Metastable Manganese Austenitic Steels

The deformation-induced $\gamma \rightarrow \varepsilon$ transformation undergoes an intermediate stage of stacking fault formation. The orientation relationship between the γ and ε phases are as follows (36) :

$$(1\bar{1}1)\gamma \,||\, (0001)\varepsilon$$

$$[110]\gamma \,||\, [11\bar{2}0]\varepsilon$$

The martensite α_m is produced by the reaction $\gamma \rightarrow \alpha_m$ in low manganese austenites, which have a high stacking fault energy, and according to the intermediate reactions $\gamma \rightarrow$ S.F. $\rightarrow \varepsilon \rightarrow \alpha_m$ in the austenites of higher manganese content, which have a lower stacking fault energy. The orientation relationship between the austenite and the α_m phase correspond to Kurdjumov's formulae (36).

$$(111)\gamma \,||\, (101)\alpha_m$$

$$[\bar{1}11]\gamma \,||\, [1\bar{1}1]\alpha_m$$

The metastable austenites can be divided into two groups according to the phase induced by work-hardening, (austenites with high manganese content and low carbon content, which show martensite ε in the work-hardened state), and austenites with low manganese content and high carbon content, which show martensite α_m in the work-hardened state.

The $\gamma + \varepsilon$ structures in the work-hardened state are interesting as they are non-magnetic and have a good weldability. The $\gamma + \alpha_m$ structures in the work-hardened state, are interesting for their wear and abrasion resistance.

The technical literature does not appear to contain any studies of the kinetics of the $\gamma \rightarrow \varepsilon$ transformation during deformation ; on the other hand the kinetics of the $\gamma \rightarrow \alpha_m$ transformation have been the subject of several publications (9) (10) (21).

The volume of the α_m phase increases slowly at first when deformation is applied. The increase in volume of the martensite produces strains in the material as a consequence of the difference in density between the

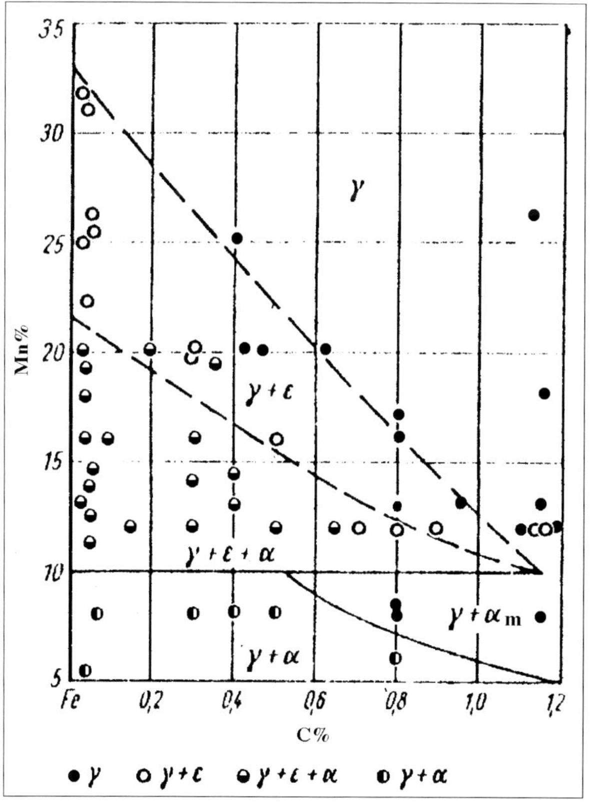

Figure 29 : Fe-Mn-C Diagram after Plastic Deformation at Room Tempature

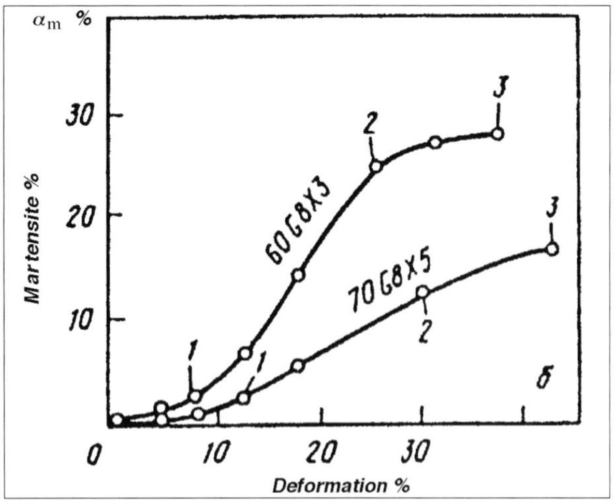

Figure 30 : Relationship between the Deformation-Induced Volume α' and Rate of Deformation of Metastable Manganese Austenites (Influence of Chemical Composition)(37)
60G8X3 0.6 C% 8.0 Mn% 3.0 Cr%
70G8X5 0.7 C% 8.0 Mn% 5.0 Cr%

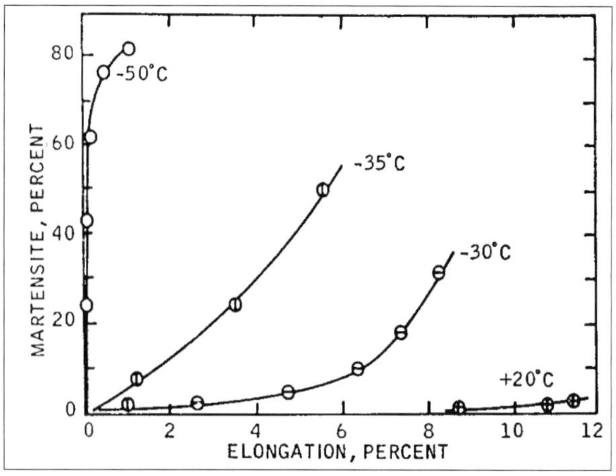

Figure 31 : Effect of Temperature and Deformation on % Martensite of a Metastable Austenite - 0.78% C - 6.16% Mn (38)

austenite and the α_m phase. This induces a martensitic transformation and the percentage of martensite increases rapidly. The austenite volume and the nucleation sites of the α_m phase decrease with the increase in the percentage of martensite, and the transformation reaches a saturation point (figure 30). The kinetics of the transformation and the percentage of martensite formed at saturation depend on the instability of the austenite and on the deformation temperature (figures 30-31). The transformation kinetics and the percentage of martensite at saturation both increase when the temperature decreases.

The kinetics of the martensitic transformation of metastable austenites is well-described by an auto-catalytic asymptotic model which takes into consideration not the transformed volume fraction VFM, but the transformed volume fraction compared to volume fraction at saturation VFM / V∞ (39) (40). The relationship is given by :

$$\frac{VFM}{V\infty} = (1 + \varepsilon^{-B/A}) - 1$$

where VFM = γ % transformed
V∞ = γ % transformed at saturation
B = constant
A = constant

A is a parameter depending on the temperature and B is a characteristic parameter of the auto-catalytic reaction related to the chemical composition. The greater the value of A, the more rapidly saturation is reached, and for the same value of A, the lower the value of B, the more unstable the material.

The presence of induced martensite in Hadfield steels is a much-debated subject. Martensite has been rarely observed with an optical microscope (36) and rarely identified by means of X-rays (36) (41) (42). Authors using the same methods of investigation, plus Mössbauer spectrometry, have all failed to reveal its presence (28) (30) (31) (33). The most commonly held opinion is that deformation does not change the austenitic structure of Hadfield steel and that cases where the presence of induced martensite was detected, were caused by decarburization or by the segregation of gamma-forming elements. The objection is pertinent and can be applied to certain observations. However detailed study of magnetic properties shows that Hadfield steel becomes ferromagnetic after work-hardening. Comparison of magnetic properties in the work-hardened state with those of a metastable austenite (13Mn-0.5C) treated under different conditions to obtain a ferromagnetic state - by cooling at low temperatures below Ms, deformation at room temperature and at 100°C - enabled some special properties to be measured, and in particular those of the magnetic phase formed during work-hardening of Hadfield steel (43) (44) (45).

The magnetic hysteresis cycle of ferromagnetic material presents two important characteristics ; the first being the intensity of magnetization at saturation (σ_s), which is dependent solely on the volume percentage of the magnetic phase and its chemical composition ; the second being the coercive field (H_c) which is dependent on the morphology and on the dispersion density of the magnetic phase.

The 13Mn-0.5C steel undergoes a martensitic transformation during cooling in the range -55°C to -155°C. An isothermal treatment in this temperature

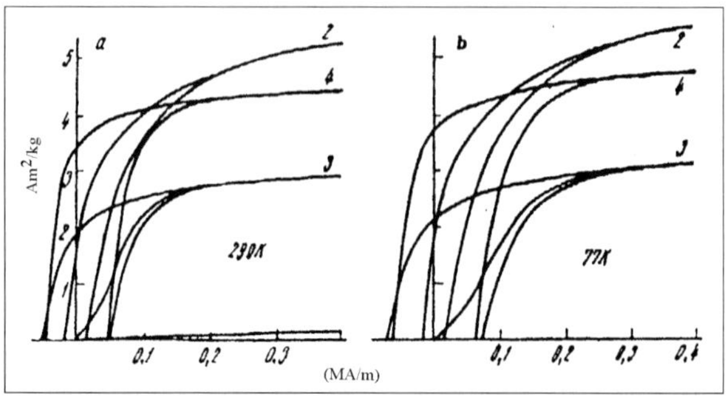

Figure 32 : Field Dependance of specific Magnetization of Specimens of 50G13 Steel at Temperature 290 K (a) and 77 K (b) : lower Curve - Initial Quenched State ; 2 - Deformation at 20°C, ε = 15% ; 3 - 100°C, ε = 60% ; 4 - 100°C, ε = 67% (43)

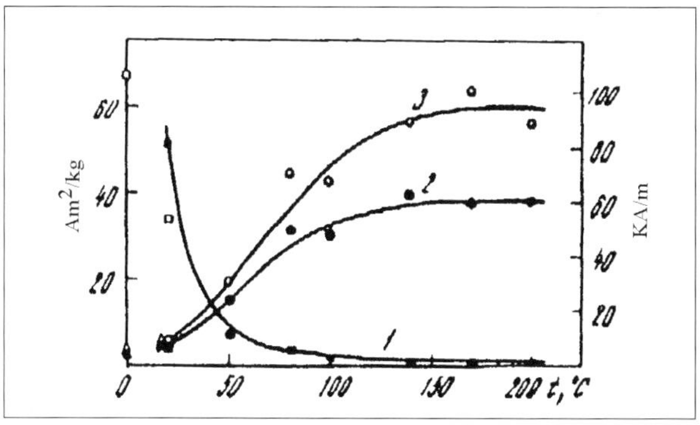

Figure 33 : Change of Specific Magnetization σ of 50G13 Steel Measured in Field 0.8 MA/m at 290 K *(1)* and Coercive Field Hc at 290 K *(2)* and 77 K *(3)* as Function of Deformation Temperature : Δ, ▲, ❐ - ε = 22% ; ❍, ●, ■ - ε = 40-50% (43)

range causes, during reheating, an isothermal martensitic transformation which prevents any subsequent transformation. The deformation and the applied strain induce the formation of martensite in this steel, and stabilize the austenite. After 15% deformation at 20°C or 60% deformation at 100°C, cooling to -196°C does not cause a martensitic

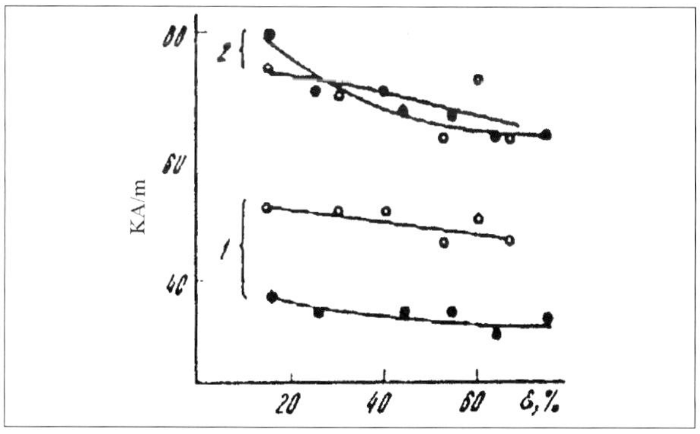

Figure 34 : Change of Coercive Force of Steel 50G13 Deformed at 100°C and of Steel 120G13, 20°C, as Function of Degree of Deformation. Measurements taken at T = 290 K *(1)* and 77 K *(2)* ; ❍ - 50G13 ; ● - 120G13

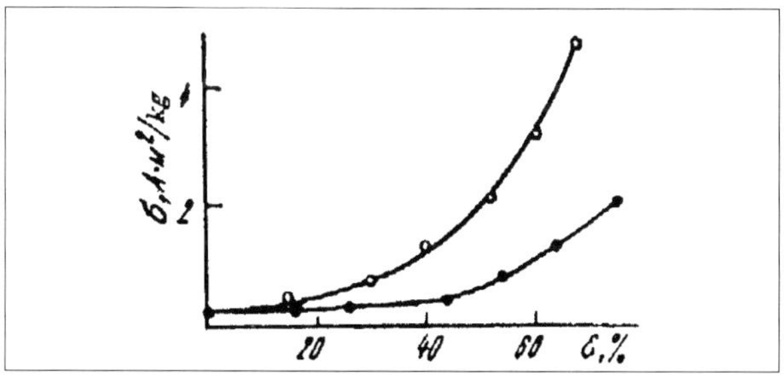

Figure 35 : Change of Specific Magnetization Measured in Field 0.8 MA/m of Steel 50G13, Deformed at 100°C and Steel 120G13, Deformed at 20°C, as Function of Degree of Deformation. Measurements taken at 290 K ; ❍ - 50G13 ; ● - 120G13 (43)

transformation. The magnetic properties of the martensite in the quenched state at low temperature or in the work-hardened state, can be measured at low temperature without the disturbance of a new martensitic transformation.

The hysteresis loop of this steel, work-hardened at room temperature, is narrow, (H_c is low). After deformation at 100°C, the cycle is rectangular in shape, with a high value for the coercive field (figure 32). The hysteresis loop after cooling below point Ms, is of the same shape as that found after deformation at room temperature, the cycle shape being especially dependent on deformation temperature. The coercive field increases, and the intensity of magnetization at saturation decreases with increase in deformation temperature, and the two variables tend towards an asymptotic value of around 100°C (figure 33).

The hysteresis loop of Hadfield steel deformed at room temperature is similar to that of metastable steel deformed at 100°C, and values of the respective coercive fields are comparable (figure 34). The loop differs solely by a lower magnetization intensity at saturation, that is, a lower volume fraction of the magnetic phase (figure 35).

Martensite, induced through work-hardening of metastable austenite at room temperature or obtained after cooling below the temperature of Ms point, is in the form of platelets, which are easy to observe under the optical microscope. The much higher coercivivity value of the martensite induced by deformation at 100°C, and the still higher value of Hadfield steel indicates that the morphology of the martensite is different, and in particular that the martensite crystals of the latter steel are of much smaller size. The value of exchange anisotropy energy (relation between the critical size of a mono-domain, and its demagnetizing field) measured for a metastable steel martensite obtained after cooling below the Ms value, or by work-hardening at 20°C, is zero. On the other hand, the exchange energy of the martensite induced at 100°C and of the Hadfield steel is exceptionally high. Assuming that the martensite crystals are the size of an elementary magnetic domain, their dimension could be taken as being of the order of 10^{-6} mm. The magnetic transformation may take place in the slip planes as already suggested by V.N. Krivobok (46) and P. Chévenard (47).

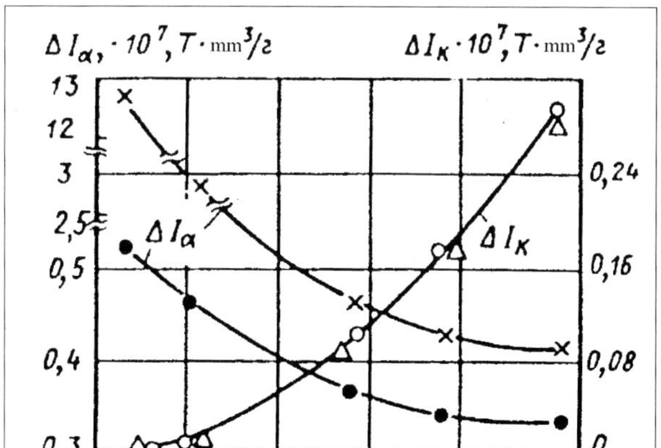

Figure 36 : Effect of C Content on Intesity of Magnetization at Saturation ΔIα and ΔIκ due to Volume Fractions of Martensite and Carbides formed after Deformation (45)
❍ ● Deformation at 20°C ; Δ X = Deformation at -80°C
Basic Chemical Composition : 14.26-15.8% Mn 0.36-0.76% Si

Measurement of the magnetization intensity of Hadfield steel at a temperature higher (20°C) as well as lower (-80°C) than the (ferromagnetic) Curie temperature of the carbide $(FeMn)_3C$, has enabled precipitation during work-hardening to be detected. Figure 36 shows that the deformation simultaneously induces a martensitic transformation and a precipitation of carbide $(FeMn)_3C$ into manganese austenites when the carbon content is higher than C = 0.8-0.96%. The percentage of martensite decreases and the percentage of carbide increases, with increase in the carbon content of the austenite (45) (figure 36). As shown by figure 35, a large deformation (30-40%) is needed to cause increase in the volume fraction of the ferromagnetic phases, and can only play a role in work-hardening at high deformation.

d) Work-Hardening Coefficient

The curve log σ - log ε is not linear, the work-hardening coefficient being very high at deformations lower than $\varepsilon = 0.001$, passing through a minimal value of around $\varepsilon = 0.02$, then increasing to reach a high value at the asymptote of the curve (figure 37). A semi-elastic behaviour may explain the initial value of the work-hardening coefficient, and its decrease is probably due to several factors : easy slip in the appropriately oriented grains, planar displacements of dislocations, like those observed in material with low stacking fault energy, and reorientation of free surfaces associated with the formation of an orange peel-like surface texture (29).

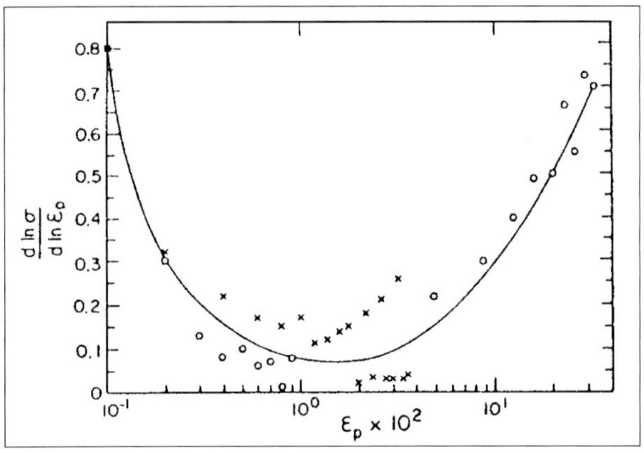

Figure 37 : Plot ot the Strain-Hardening Exponent *vs* Plastic Strain (29)
(The two different kinds of symbol stand for two specimens.)
C% 1.25, Mn% 13.1, Si% 0.47

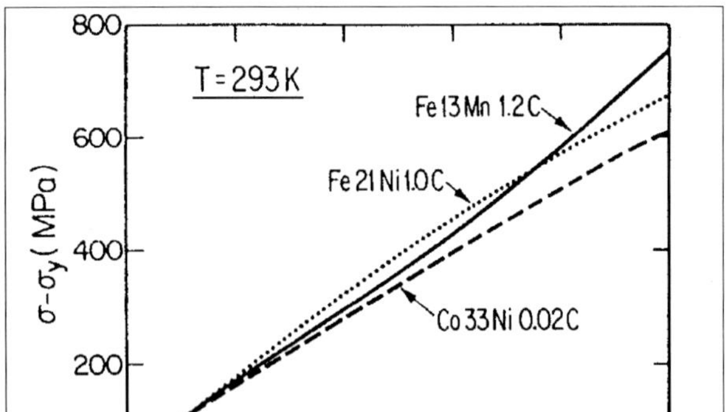

Figure 38 : True Tensile Stress minus Yield Stress *vs* True Plastic Strain Curve Comparing the Room Temperature Work-hardening Behaviour of Hadfield Maganese Steel (25) (Solid Line), Fe-21Ni-1/0C (dotted line) and Co-33Ni-0.02C° (dashed line)

Comparison of the work-hardening curve $(\sigma-\sigma_y)$-ε of a Hadfield steel (13% Mn - 1.2% C) with that of austenites which are deformed either by slip (Fe - 20% Ni - 1.04% C) or twinning (Co - 33% Ni - 0.02% C) enables the effect of the two deformation processes on the work-hardening of Hadfield steel under tensile test at room temperature to be evaluated (figure 38).
The three austenites under consideration have the same yield strength and a similar behaviour at low deformations ($\varepsilon \leq 0.05$). The curve of the Hadfield steel is situated at deformations between $0.06 < \varepsilon < 0.22$, between that of the austenite deformed by twinning, which work-hardens the least, and that of the austenite which work-hardens by slip, and which work-hardens the most. The curves bend at a deformation of the order of 0.2, that of the Hadfield steel turning upwards, and that of the other two turning downwards. The result leads to a crossing of the Hadfield curve with the curve of the austenite of Fe-Ni-C, at a deformation of the order of 0.22, and to a greater work-hardening of the Hadfield steel at higher deformations.

The work-hardening curves of the Hadfield steel and of the Co-Ni austenite are not directly comparable : although the two materials have the same twinning kinetics (figure 23), the grain size and twin volume of the Hadfield steel are coarser. The increase in resistance, due to the twinning $\Delta\sigma_S$ is given by the Remy relation (48) :

$$\Delta\sigma_S = K_T(2t)^{-1}\left(\frac{f}{1-f}\right)$$

where K_T = constant
t = thickness of the twins
f = volume fraction of the twins

It can be calculated that the work-hardening curve of a Co-Ni austenite, which would have twins of the same thickness as the Hadfield steel under consideration, would be displaced towards values about 35% lower. In order to explain the strong effect of twinning on the work-hardening of Hadfield steel, Adler and colleagues (25) put forward the hypothesis that twinning may not produce real twins. Shearing during twinning would convert the octahedral sites occupied by the carbon atoms into tetrahedral sites where they would be trapped, unless they have diffused away. The occupation of tetrahedral sites, in solid solution where the repartition of carbon atoms is not random, would cause a distortion of the network, giving a centred orthorhombic structure.

These distortions would produce a hardening analogous to the distortion of tetragonal martensite. Comparison of the effect of the volume fraction of twins in a Hadfield steel, and the effect of volume fraction of martensite of a metastable austenite on work-hardening, confirms that the volume fractions of twins and of martensite have a comparable quantitative effect (figure 39). Remy's equation should be modified by an additive term $\Delta\sigma_0 f$, proportional at a first approximation to the twin volume fraction, in order to take into account the specific ratio of the pseudo-twinning phenomenon. The numerical value of $\Delta\sigma_0$ is of the order of 620 MPa.

By referring to the work-hardening of FeNiC and

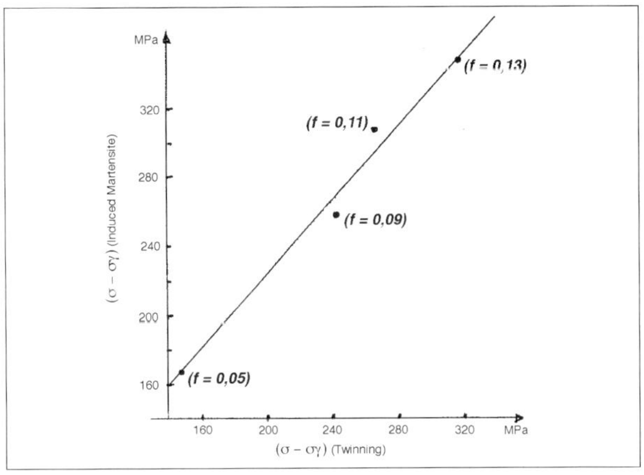

Figure 39 : Comparison of the Effect of Twinned Volume Fraction and Martensite Volume Fraction Induced by the Work-hardening of Manganese Austenites Induced martensite effect Twinnig effect
Hadfield steel C% 1.18, Mn% 12.9, Si% 0.2
Metastable austenite C% 0.98, Mn% 4.1, Si% 0.57

CoNi austenites, it can be ascertained that twinning provides work-hardening comparable to that of Hadfield steel up to a deformation of the order of 0.10, and slip to a deformation of the order of 0.22. The work-hardening of Hadfield steel becomes greater with higher deformation. The predominance of one of the two deformation modes depends on chemical composition and temperature which regulate the stacking fault energy (figure 40). Pseudo-

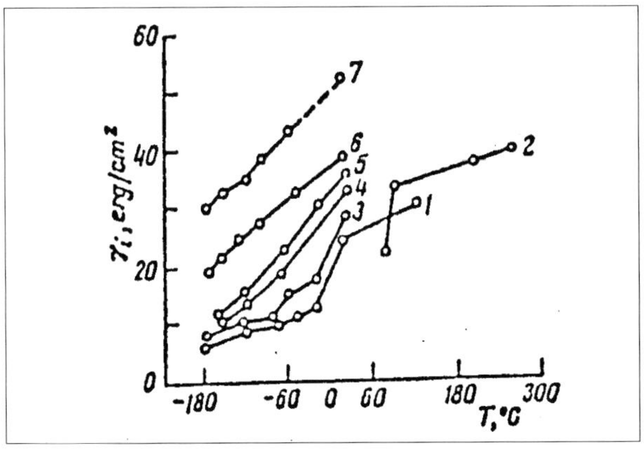

Figure 40a : Change of Austenite Stacking Fault Energy *vs* Temperature for Different Alloys of Iron, Manganese and Carbon (49)
Average Grain Size = 10-15 μm
1) Fe-17.8 Mn-0.47 C
2) Fe-13.7 Mn-0.23 C 3) Fe-14.9 Mn-0.35 C
4) Fe-12.2 Mn-0.5 C 5) Fe-11.8 Mn-0.65 C
6) Fe-12 Mn-0.76 C 7) Fe-13.2 Mn-1.35 C

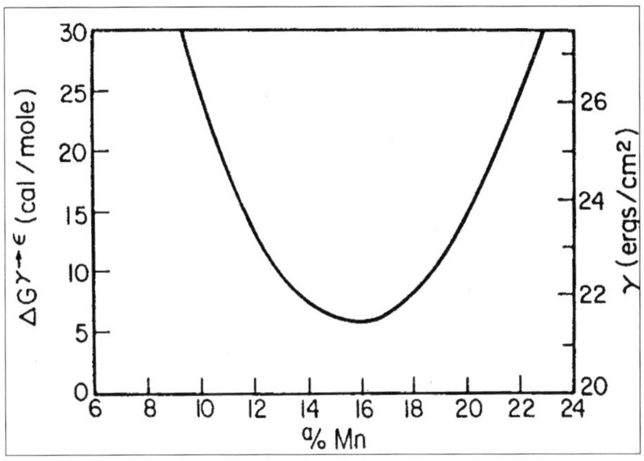

Figure 40b : Approximate fcc → hcp Free Energy Difference. $\Delta G\gamma \rightarrow \varepsilon$. and Stacking Fault Energies. γI-*vs* Mn Content. for a Constant Carbon Content (C = 1.15%) (25)

twinning may explain the work-hardening of low stacking fault energy austenites, and does not explain the work-hardening of high stacking fault energy austenites, which are deformed by slip. As the work-hardening of these austenites seems to depend little on the type of deformation, twinning or slip, pseudo-twinning and dynamic strain-aging would be equivalent.

e) *Effect of deformation conditions on work-hardening*

1. Deformation temperature and deformation rate

The tensile strength and elongation of Hadfield steel passes through a minimum value in the range 150-200°C (figure 41). Elongations are uniform from below room temperature to about 300°C. Fracture is accompanied by a reduction in area at higher

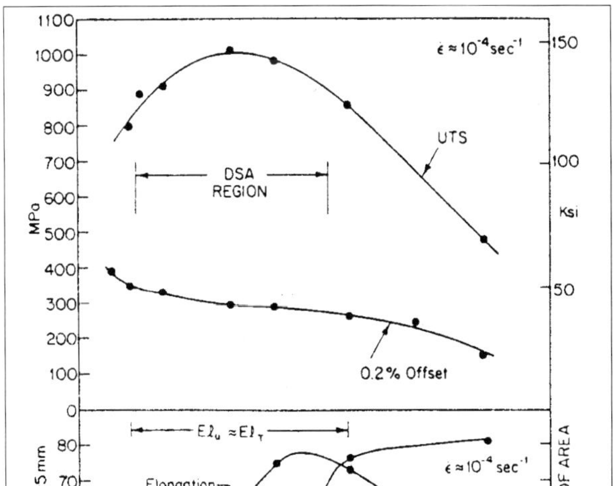

Figure 41 : Temperature Dependence àf Tensile Properties of Hadfield Steel (El_u = uniform elongation, El_T = total elongation) (31)

temperatures. The work-hardening stress ($\sigma-\sigma_\gamma$) corresponding to a given deformation rate, is almost constant between room temperature and about 300°C. The work-hardening stress decreases and dynamic strain-aging disappears, at higher temperatures. M_3C carbides precipitate above 400°C, and the work-hardening stress increases once more.

The work-hardening stress increases below -50°C, where dynamic strain-aging does not occur, which might be explained by the twinning which is encouraged by low temperatures (figure 42).

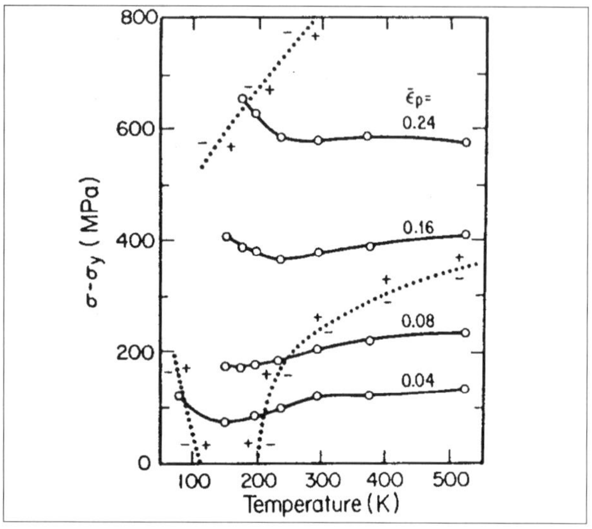

Figure 42 : Tensile Flow Stress minus Yield Stress *vs* Temperature for True Plastic Strains of 0.04, 0.08, 0.16 and 0.24 (Dashed curves depict the positions of inflection points in the true stress-true plastic strain curves and, thus, separate regions of upward curvatur e ($d^2\sigma/d\varepsilon_p^2 > 0$, indicated by +) and downward curvature $d^2\sigma/d\varepsilon_p^2 < 0$, indicated by -) (25)

Increase in deformation rate from $10^{-4}s^{-1}$ to $10^{-3}s^{-1}$ does not seem to have an effect on twinning rates, and the appearance of the serrations of the stress-strain curve is altered, changing from type B to type A (figure 43). Simultaneously, in the domain of dynamic strain-aging temperature the average deformation stress is lowered.

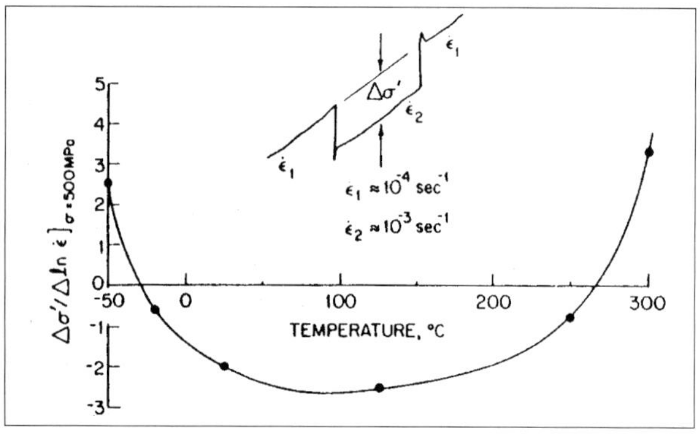

Figure 43 : Variation of Strain-Rate Dependence of Flow Stress with Temperature in Hadfield Steel (31)

2. Deformation mode

Work-hardening of Hadfield steel is greater under compression than under tension (figure 44). The twinning rate being independent of deformation mode, this difference may be attributed to the textures developed, texture [111] under tension and texture [110] under compression. The Taylor factor, which is calculated for symmetrical deformations of slipping and simultaneous twinning, shown in table 7 for $\alpha<1$, is greater for a compression texture [110] than for a tension texture [111] (25).

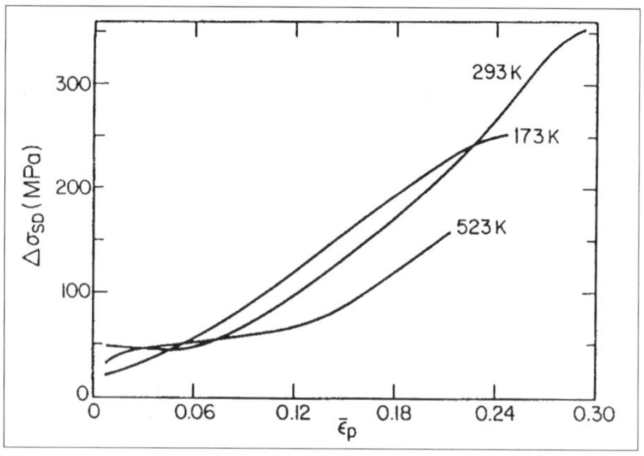

Figure 44 : $\Delta\sigma_{SD}$ *vs* Equivalent True Plastic Strain Curves for 173 K, 293 K, and 523 K ($\Delta\sigma_{SD}$ = (σ compression - σ tension) (25)

The elongation of Hadfield steel under tension is large and macroscopically uniform to fracture which occurs without reduction in area, and it presents a considerable advantage over most other steels.

The absence of area reduction is due to the fact that elongation during deformation is always less than the work-hardening coefficient. Several physical models have been suggested to explain this : incipient area reduction, in the domain of dynamic work-hardening, might be caused by a local impoverishment of interstitials in solution, they might become insufficient to block mobile dislocations, and would thus enable a localized plastic deformation to develop and to cause area reduction. The high carbon content of Hadfield steel would not permit this process (50). The process of area reduction would be inhibited by the formation of stacking faults, according to a

Tension	**M[100]**	**M[111]**	**M[111]/M[100]**
$\alpha>2/\sqrt{3}$; slip	2.449	3.674	1.50
α = 1; slip and twinning	2.449	3.182	1.30
α = 2/3; slip and twinning	2.449	2.121	0.87
$\alpha = 1\sqrt{3}$; twinning	2.449	1.837	0.75
Compression	**M[100]**	**M[110]**	**M[110]/M[100]**
$\alpha>2/\sqrt{3}$; slip	2.449	3.674	1.50
α = 1; slip and twinning	2.121	3.510	1.65
α = 2/3; slip and twinning	1.414	3.157	2.23
$\alpha = 1\sqrt{3}$; twinning	1.225	3.062	2.50

Table 7 : Taylor Factors for Deformation by Combined Slip and Twinning (25)

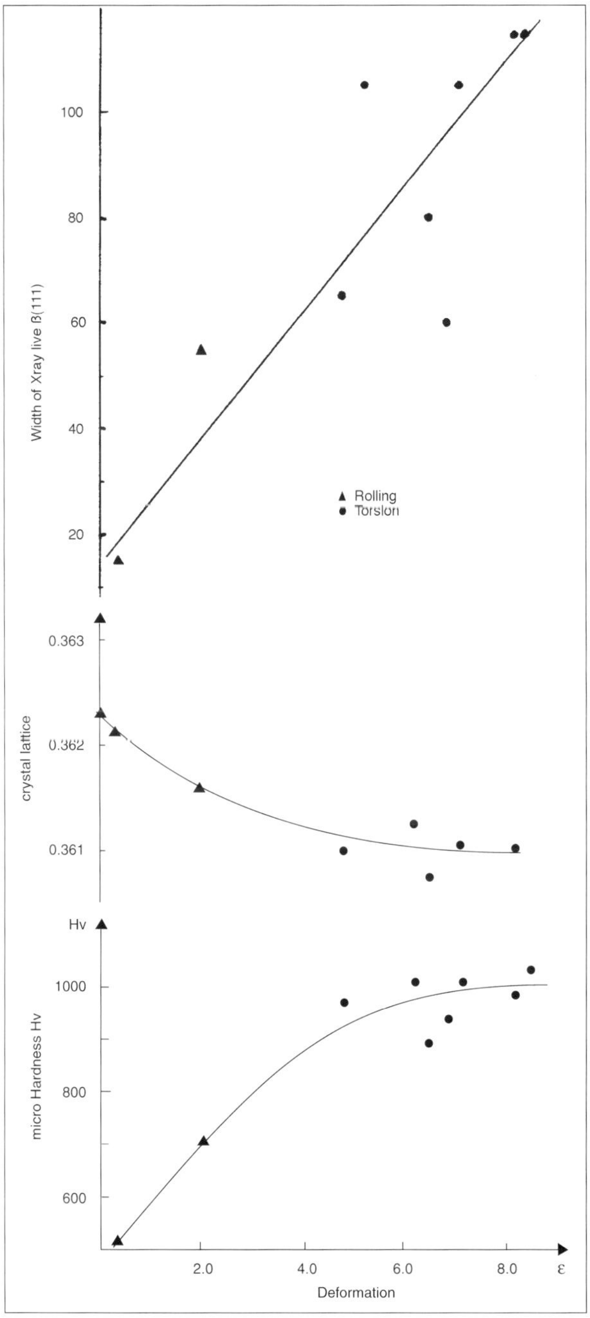

Figure 45 : Effect of Shearing Deformation on Micro-Hardness, Crystal Lattice and Width of X Ray Line for Austenite $\beta_{(111)}$ of Hadfield Steel (C% 1.2 - Mn% 13.4)[28]

process similar to the one observed in steel SS301, or the area reduction may be inhibited by martensite formation (51).

The capacity for deformation under tension is less than for compression, nevertheless the latter deformation process gives a work-hardening rate which is much less than those observed in process in the surface layers. Fracture deformation in torsion is considerably greater than in compression, and torsion tests produce hardness levels comparable to that of work-hardened surface layers. Figure 45 shows that the micro-hardness of Hadfield steel and the crystal lattice of austenite tend towards a limiting value beyond a torsion deformation value of the order of

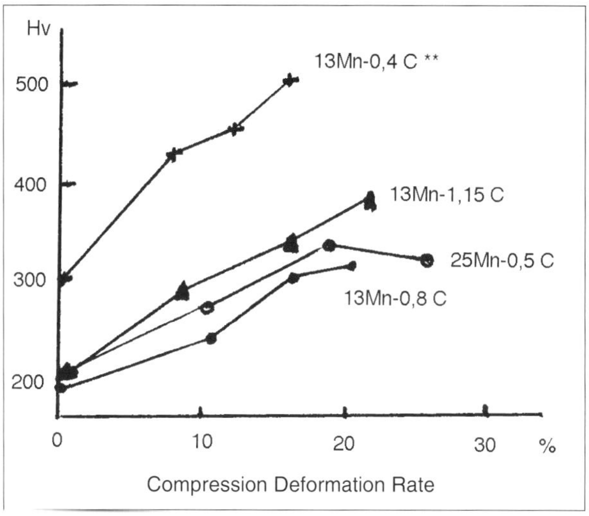

Figure 46 : Effect of Compression Deformation Rate on the Increase in Hardness of Metastable Austenites, and of Hadfield Steel (30)
* 10% ε at the Quenched State. Compression-deformed at -196°C, 20% ε
** Traces α_m at the Quenched State. Compression-deformed at Room temperature, 20% α_m

$\varepsilon = 5$. The widening of the austenite line ß(111) is proportional to the deformation, which implies an increase in the number of crystal defects, although the hardness reaches a saturation value.

f) Metastable austenites

Figure 46 enables a comparison to be made of the hardening after work-hardening of the metastable austenites $\gamma \rightarrow \varepsilon$ and $\gamma \rightarrow \alpha_m$ with that of Hadfield steel. The steel with 13Mn, 0.8C, metastable at deformations higher than the applied ones, is austenitic after deformation, and the steel with 25Mn, 0.5C, dual-phase (γ + 10%ε) before deformation, presents twice as much ε martensite after deformation at -190°C. These two steels work-harden less than the Hadfield steel. The steel with 13Mn, 0.4C, which has an Ms value near room temperature, presents 20% α_m induced martensite after deformation at room temperature and work-hardens more than the Hadfield steel. The difference in work-hardenability of these austenites can be attributed to the kinetics of twinning and of martensitic transformations.
The study of metastable austenites ($\gamma \rightarrow \varepsilon$), used for their non-magnetic properties, has concentrated in particular on yield strength, and the study of work-hardening mechanisms has not been extensive. On the other hand the work-hardening of metastable austenites ($\gamma \rightarrow \alpha_m$) has been the object of several studies.

α_m martensite is induced in austenites with an Ms value above room temperature, as soon as a deformation is applied. The stress-strain curve of these austenites shows serrations due to the martensitic transformation from the beginning of deformation until fracture. The work-hardening coefficient is high, and there is little elongation (2 to

Chemical Composition %				M_S, % Martensite and Deformation at Fracture				Coefficients of equation : $\sigma = A\varepsilon^a Exp(-\alpha\varepsilon^\beta) + M\varepsilon^m[1-Exp(-\alpha\varepsilon^\beta)]$					
C	Si	Mn	Mo	M_S, (°C)*	% Mar -tensite (Quenched state)	ε Frac -ture	% Induced Martensite	A (MPa)	a	α	β	M (MPa)	m
0.68	0.62	5.70	0.99	90	17.9	0.018	7.6	840	0.145	54	1.97	9710	0.169
0.98	0.57	4.10		64	4.5	0.048	12.8	650	0.127	5.9	1.32	3640	0.18
1.12	0.59	3.02	1.02	51	7.4	0.024	6.2	610	0.092	16.6	1.6	4210	0.13
1.13	0.57	4.04	1.00	10	0	0.049	1.5	830	0.141	17.0	0.99	2070	0.50
0.71	0.57	5.62		95	18.6	0.016	4.8	840	0.136	68	1.78	2000	0.038
1.12	0.61	4.06		23	0	0.107	1.0	5800	0.12				
1.13	0.60	5.80		-37	0	0.138	0.01	1100	0.26				

Table 8 : Ms Temperature, % Martensite, and Coefficient of Borik's Equation for several Metastable Austenites. *Calculated values

5%). An empirical relation has been suggested for calculating the stress-strain curve of two-phase metastable steels (52) :

$$\sigma = A\varepsilon^a Exp(-\alpha\varepsilon^\beta) + M\varepsilon^m[1 - Exp(-\alpha\varepsilon^\beta)]$$

where σ is the applied stress, ε the strain, the terms $Exp(-\alpha\varepsilon^\beta)$ and $[1 - Exp(-\alpha\varepsilon^\beta)]$ giving respectively, according to the autocatalytic law for the fractions of induced martensite and untransformed austenite, the terms $A\varepsilon^a$ and $M\varepsilon^m$, as work-hardening stress of the austenite and of the martensite.
The equation coefficients are shown for a few metastable steels, in table 8.
The stress-strain curve of metastable austenitic steels of type 6.1, can be practically superimposed over that of the Hadfield steel. It shows serrations due to dynamic strain-aging, and differs in that they have lower fracture elongation, which is nevertheless higher than that of two-phase steels, at 10 to 15%. Deformation occurs by slip, as twinning is inhibited in these low manganese austenites by a high stacking fault energy (figure 40). Tensile fracture of these steels is premature, due to the embrittling effect of induced martensite when a critical strain is reached. These steels do not have any advantage over Hadfield steel in uniaxial tension. Their greater work-hardenability can be used only in deformation systems which permit strain higher than the critical strain for uniaxial tension.

The curve log σ - log ε of a metastable austenite formed in a steel of composition 1.08% C - 6% Mn - 0.22% Mo, obtained during static compression, shows a change in slope after deformation of the order of ε = 0.12, which practically corresponds to the elongations in tensile fracture (figure 47). The deformation corresponds to a change in slope of the compression curve, and correlates with the occurrence of ferromagnetic properties which then rapidly increase with an increase in deformation (figure 48). The change in slope of the compression curve corresponds to the contribution of the martensite induced by work-hardening.

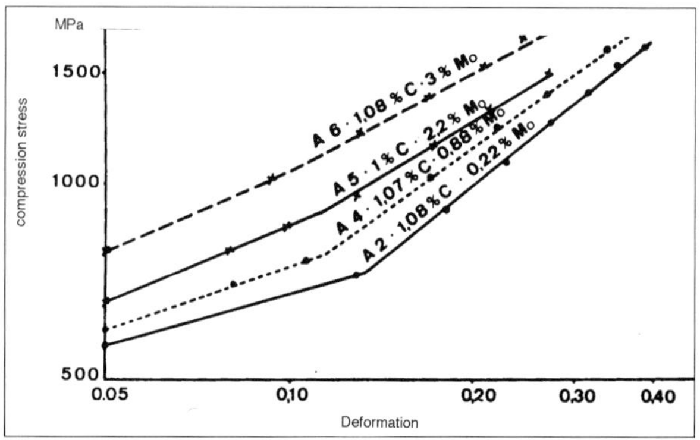

Figure 47 : Stress-Static Deformation Curves of Compression of Metastable Austenites with 6% Mn and Different Mo Contents

The addition of molybdenum to these steels lowers the critical deformation of martensite induction, and increases the volume fraction of induced martensite, due to an increase of the deformation rate (figures 47-48).

The volume fraction of induced martensite depends on the deformation rate. A drop test in compression shows that deformation work is greater at an impact rate of 7.0 ms^{-1} than at 3.5 ms^{-1} (figure 49). This observation suggests that the volume fraction of the the induced martensite, in other words, the instability of the austenite, depends on the deformation rate.

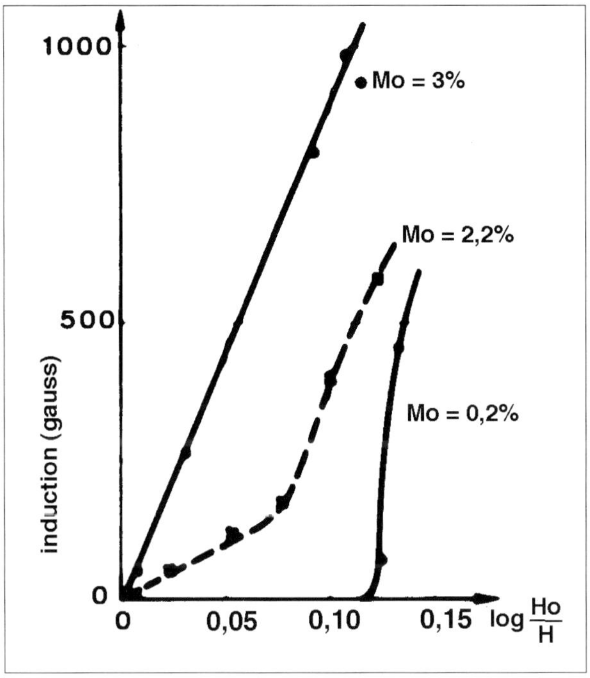

Figure 48 : Relationship between Magnetic Induction and Compressive Deformation of Metastable Austenites with 6% Mn - 1% C and different Mo Contents

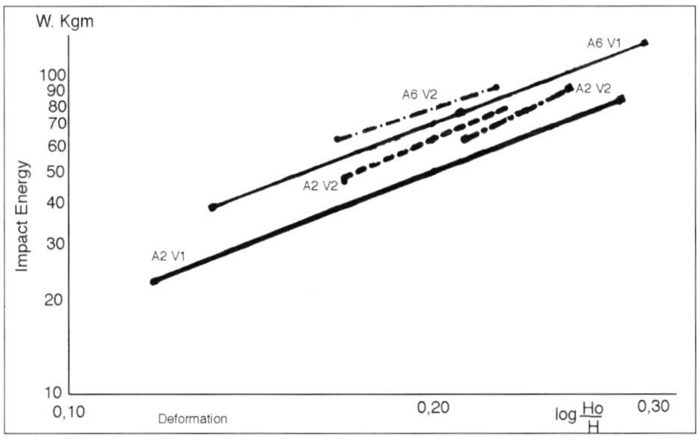

Figure 49 : Relationship between Impact Energy and Compressive-Deformation of Metastable Austenite - Effect of Deformation Rate and Chemical Composition

Rate of Impact V_1 3.5 ms^{-1}, V_2 7.0 ms^{-1}

Chemical composition A2 1.08 C% 6.0 Mn% 0.22 Mo%
A6 1.08 C% 6.0 Mn% 3.00 Mo%

g) Static strain-aging of Hadfield steels

The static strain-aging of Hadfield steel has been studied after a deformation of 4% at -25°C, to avoid dynamic strain-aging. The deformation stress undergoes a small increase ($\Delta\sigma$ = 15MPa) independent of temperature, and of the aging time at temperatures between 22 and 200°C., during periods of less than 2 min. and up to 7 days. Strain-aging is greater after prolonged aging at a temperature between 300 and 400°C (figure 50). An increase in the concentration of carbon atom clusters after strain-aging has been observed by investigators, and characterizes a 'pre-precipitation' state of carbides (3%). Others have noticed a precipitation of carbide $(FeMn)_3C$ using the optical microscope (31). The apparent activation energy of the strain-aging process corresponds well with the bulk diffusion energy of the carbon. The aging after work-hardening can be attributed to the formation of Cottrell atmospheres and to the precipitation of the carbide $(FeMn)_3C$.

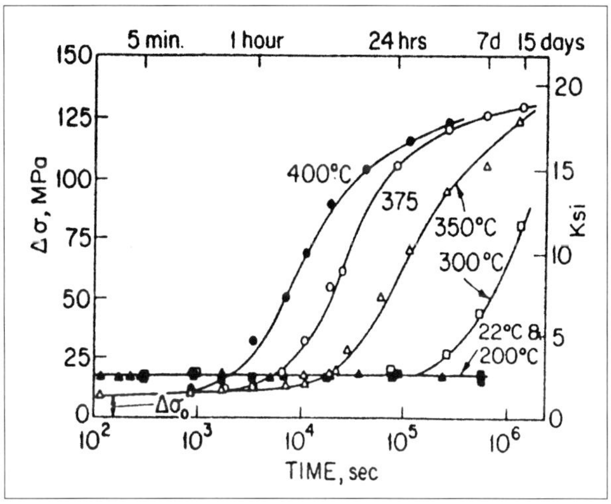

Figure 50 : Static Strain Aging of Hadfield Steel at High Temperature (31)
C% 1.13, Mn% 11.4, Si% 0.2, Ni% 0.17, Cr% 0.16, Mo% 0.08
Pre-Deformation of 4% at -25°C

h) Critical work-hardening of Hadfield steel

The recrystallization diagram of work-hardened Hadfield steel (figure 51) shows that critical work-hardening of Hadfield steel corresponds to a 10% deformation. The grain size begins to coarsen after work-hardening of 10% above a temperature of 800°C. (The holding time is not specified.)

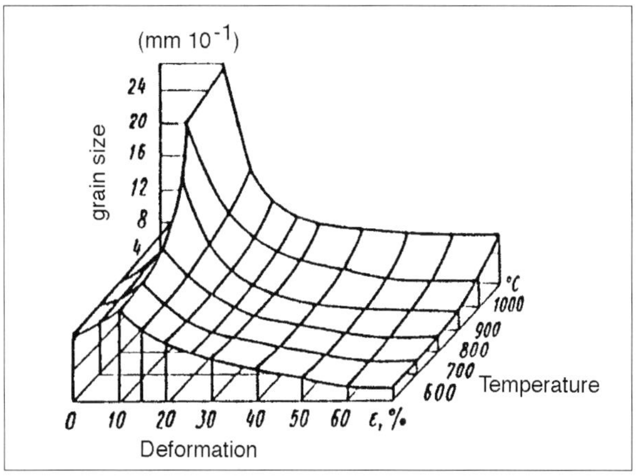

Figure 51 : Recrystallization Diagram of Hadfield Steel (53)
(Treatment Time is not specified)

III. Fracture Mechanisms

a) Brittle fracture

1. Fracture by cleavage

According to V.D. Sadowskii and colleagues (54), the fracture at -196°C of Hadfield steel with 0.014% P occurs by cleavage, in the {111} plane when it is oriented in a direction similar to that of the applied force : if the disorientation is great, the fracture is of a mixed type, being ductile and brittle.

Cleavage does not occur in the macro-twin boundaries , but probably in the fine twin boundaries, which form during deformation before cleavage. Another hypothesis, less likely, however, than the first, is supported by Mössbauer tests and the measurement of the austenite lattice, which have shown a redistribution of carbon in the austenite during deformation : this produces a weakening of the {111} plane due to localized segregation of carbon, and the occurrence, or dynamic formation, of clusters in this plane.

Transcrystalline cleavage at low temperature is typical of single crystals, but it can also be observed in appropriately oriented columnar crystals, and in polycrystals, when the latter are very pure (P ≤ 0.003%) (figure 52).

2. Intergranular fracture

Dendritic segregation of impurities such as phosphorus, considerably lowers the resistance to intergranular cohesion. The occurrence of intergranular precipitates of carbides, oxides, nitrides, or a change of phase - martensite, pearlite - in the grain boundaries, produces interfaces less resistant than cleavage, and there is a transition to intergranular fracture.

The critical content of phosphorus at which there is a transition from cleavage fracture to intergranular fracture, is very low. The toughness at low temperature of two steels, one with 0.005% P and the other with 0.014% P, is identical, and their fracture surfaces have an intergranular aspect (figure 53). However, the transition from fracture by cleavage to intergranular fracture occurs with a phosphorus content of around 0.005%. Industrial steels show practically no fracture by cleavage because of their residual phosphorus content, with the exception of columnar crystallization areas, where it may sometimes occur.

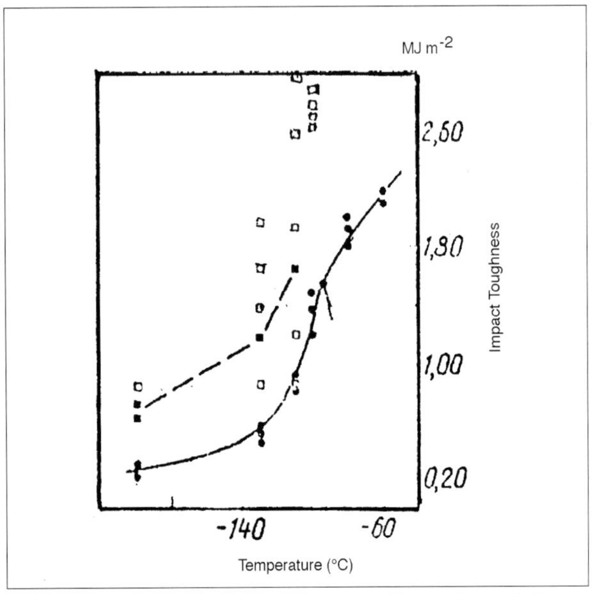

Figure 52 : Effect of Texture on Temperature/Impact Strength Curve of Hadfield Steel with 0.014% Phosphorus (54)
● polycrystalline ■ monocrystalline □ colummar

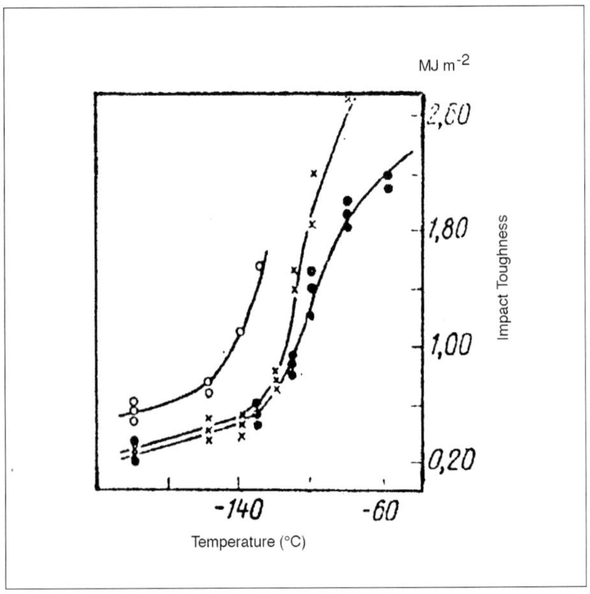

Figure 53 : Impact Strength Temperature Curve of Hadfield Steel (1/2% C, 13.2% Mn) with Different P contents (54)
❍ 0.003% X 0.005% ● 0.014%

3. Transgranular fracture

Carbide precipitation as platelets in different crystal planes has the effect of hardening the austenite, the critical stress for slip increases and the slip bands are narrower than when carbides are absent. As the carbides have no plasticity and the carbide-matrix interface has little resistance, two kinds of cracks can develop during plastic deformation, according to the orientation of the carbide platelets in the stress field. The carbide platelets oriented towards the applied force crack at the intersection of the slip lines, the width of the cracks corresponding to the thickness of the slip band, but they show no tendency to spread into the matrix, and do not seem harmful. On the other hand, when the platelets are oriented at a 45° angle to the applied force, the carbide-matrix interface shows low shearing strength, the bonding between the matrix and the carbides is destroyed. The matrix deforms in the direction of the strain, and the right-hand section of the crack takes on a lenticular shape.

The crack then spreads out from one end of the carbide to the other, in a straight line or in steps, according to whether the nearest carbide has precipitated in the same plane or in a plane at right angles to it.

Another mechanism of transgranular fracture has been observed in manganese austenitic alloys rich in aluminium (30Mn - 10Al - 1Si - 1C) and deformed at low temperature (23). The cracks have a faceted appearance, with the facets oriented towards the plane {111}. Fracture does not occur by cleavage, but by the spreading of micro-cracks into the slip planes.
In the fracture process, no twinning, dislocation ordering or phase changes accompany the deformation, which is localized exclusively in the slip bands. Very high stress concentrations occur at the intersection of the slip bands, where micro-cracks can initiate. These micro-cracks inhibit locally the formation of secondary slip, and, under the effect of a large shearing deformation, combined with stress concentrations near their extremities, these microcracks may easily spread along the slip planes, to form macro-fissures.

The fracture, by cracking, of the slip planes observed at low temperature, occurs particularly under simple stress in a ductile, non-embrittled material. The embrittling factors - triaxial stress, deformation rate, transgranular carbide precipitation - explain the occurrence of transgranular fracture which can become predominant. Brittle fracture of the 30Mn - 10Al - 1Si - 1C alloys at low temperature, occurs by cracking of the slip planes under stress, and by intergranular fracture under impact bending of notched test pieces. Stress fractures at low temperatures of work-aged alloys, show a mixture of transgranular fractures and fracture by cracking of slip planes.

b)Ductile fracture

The transition from brittle intergranular or transgranular cleavage fracture, to ductile fracture with the formation of cusps, occurs within a relatively large temperature range, of the order of 150-200°C, with the temperature being lower for high purity alloys.
In the transition area, the percentage of brittle fractures increases and the percentage of cusps decreases, with a decrease in temperature. A.P. Gulyaev and V.M. Chulkova (55) have moreover observed a particular aspect of transcrystalline ductile fracture in the transition area, in the form of undulating relief zones (elongated cusps, formed by shearing or by cleavage). The percentage of this ductile appearance peaks at 40-50%, when the intercrystalline fraction represents 15 to 30% of the fracture area (figure 54).

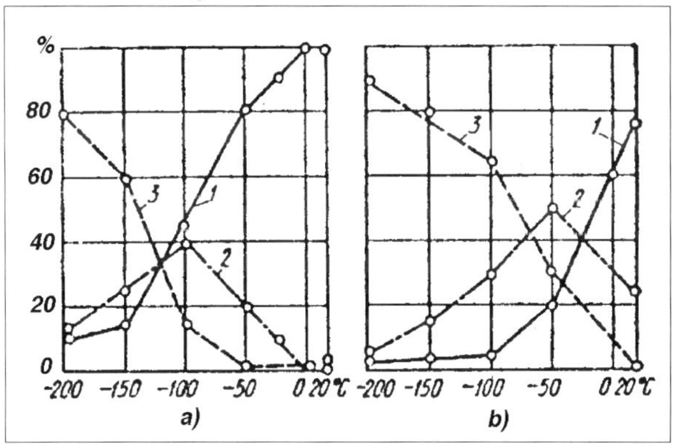

Figure 54 : Relationship between Fractographic Aspect of Fracture of Hadfield Steels and Temperature (55). a) 0.001% P ; b) 0.078% P
Curve 1 % cupels, Curve 2 % corrugated zones,
Curve 3 intergranular zones

c) Fatigue fracture

1. Cyclic work-hardening and crack initiation

Cyclic work-hardening, which precedes the initiation of fatigue cracking, is strongly dependent on the previous mechanical history of manganese austenites, which makes the comparison of data difficult. According to authors, the stress-strain deformation cycle evolves without discontinuity (56) (57) (58), the stress increases, traverses a maximum value, then decreases during a large part of the lifetime (figure 55). Cyclic deformation produces a localized plastic zone at the notch tip, at a depth of 0.2-0.3 mm., which appears in the form of slip lines and gives an increase in hardness.The number and amplitude of the slip bands as well as the hardness increase as the initiation point is approached (figure 56). The instability of the cyclic behaviour and the impossibility of directly transposing the number of cycles leading up to crack initiation and to fracture (figure 57) makes the evaluation of the initiation period relatively complicated. An elasto-plastic model from Neuber's rule :

$$\sqrt{E\Delta\sigma_{max}\,\Delta\varepsilon} = C^{te}$$

where E = Young modulus, ε = deformation, C = constant

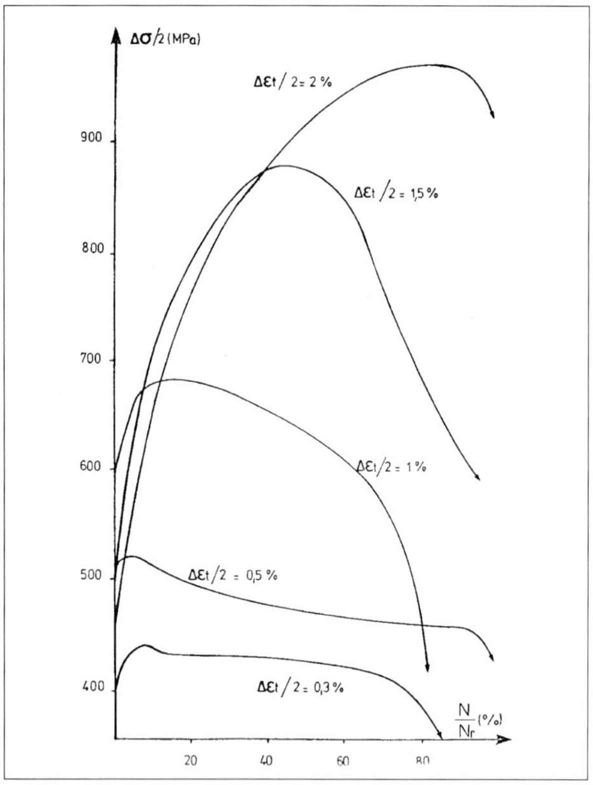

Figure 55 : Evolution of Cyclic Stress as a Function of the ratio : Nr of cycles applied / Nr of cycles at fracture. for Different Applied Cyclical Deformation Rates (58)

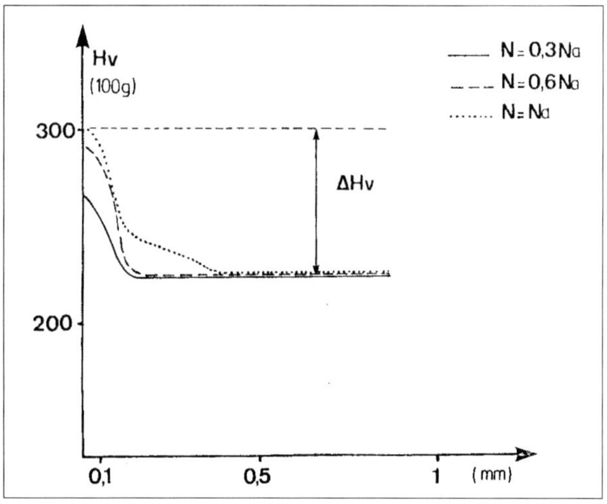

Figure 56 : Hardness at Notch Root before and at the start of Crack Propagation (58)

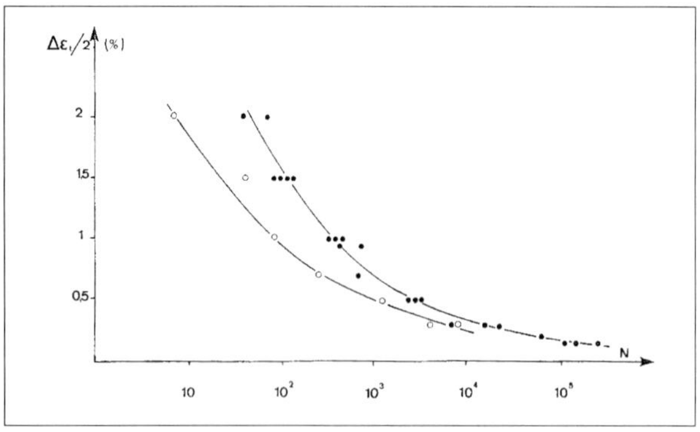

Figure 57 : Relationship Between Cyclic Deformation and the number of Cycles at Start of Propagation and at Fracture (58)
❍ N° of Cycles at Start of Propagation ● N° of Cycles at Failure

gives an estimate (figure 58). The value of the criteria can be obtained by evaluating the maximum deformation at the notch tip, according to the hardness measured there (figure 59) and the corresponding stress with the help of a stress-strain calibration curve (figure 60) established by oligocyclic imposed strain fatigue test using smooth specimens.

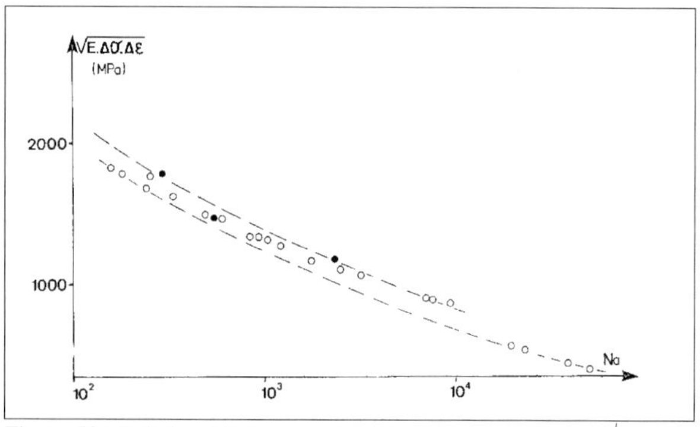

Figure 58 : Relationship between the Elasto-Plastic Criterion √EΔσΔε. and the N° of Cycles at Start of Propagation (58)
❍ √ EΔσ max Δε (modified elasto-plastic model)
● √ EΔσ' max Δε (Measurements of microhardness)

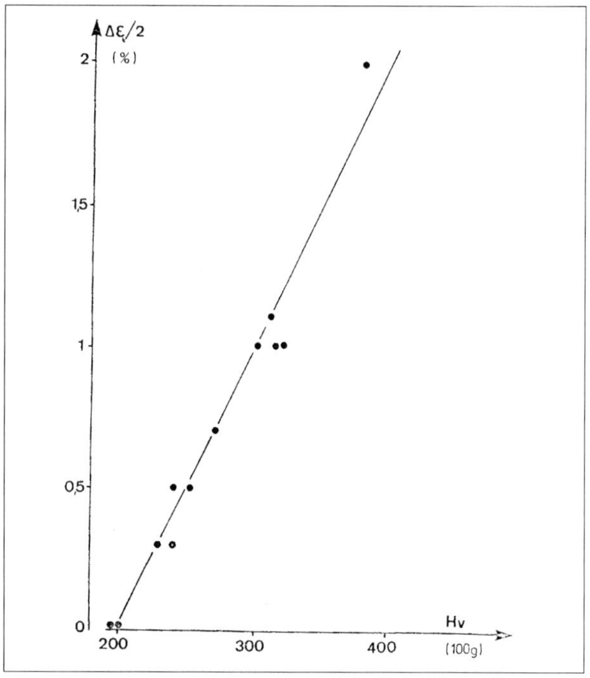

Figure 59 : Relationship between Cyclical Deformation and Hardness (58)

2. Notch sensitivity

Notch sensitivity $q = \frac{K_f - 1}{K_t - 1}$

where $K_f = \frac{\sigma_L}{\sigma_E}$ ratio fatigue limit of notched specimen/fatigue limit of smooth specimen

K_t = coefficient of stress concentration,

indicates a material's capacity to adapt to a local stress concentration which can cause fatigue

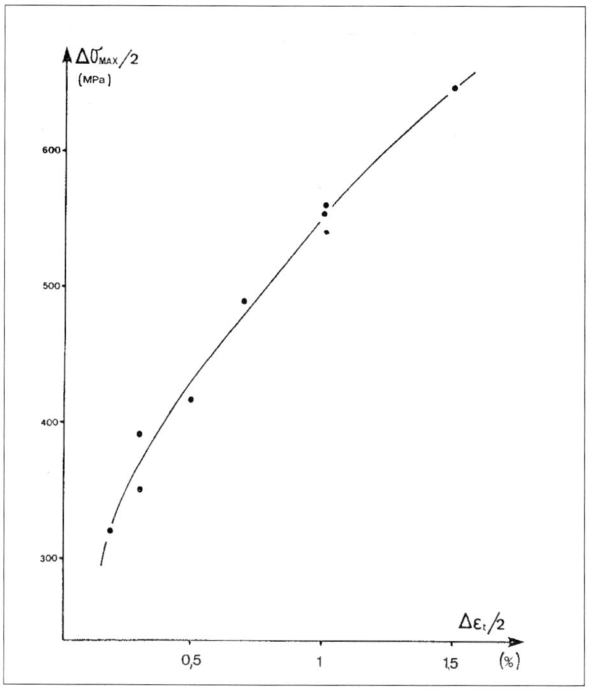

Figure 60 : Relationship between Cyclical Deformation (Measurements of Smooth Test Pieces in Controlled Deformation) (58)

cracking. For a material with no notch sensitivity, the value of *q* would be zero, and the more notch sensitive a material is, the higher the value of *q*.

The notch sensitivity of austenitic manganese steels depends on their instability. The capacity to adapt at the notch tip of an unstable austenite is reduced, and the value of *q* increases when K_t and work-hardenability increase (table 9).

Table 9 : Fatigue Characteristics of Manganese Austenitic Steels

*a. Chemical Composition % *A-F, and K = figure 61*

Casting*	C	Mn	Si	P	S	Cr	Ni	Mo	Ti	Rare Earths
A	1.25	13.1	0.47	0.030	0.008	0.13	0.1			
B	1.06	13.4	0.87	0.051	0.003	1.30		0.66	0.14	
C	1.11	12.6	0.71	0.047	0.01	1.11	0.26	0.89	0.04	
D	1.13	12.5	0.55	0.028	0.01	1.14	0.24	0.59	0.06	
E	1.22	13.2	0.49	0.025	0.005	0.30				
F	0.78	13.4	0.43	0.023	0.005	0.11		1.13		
G	1.10	13.0				1.3	1.2	0.60	0.03	
H	1.10	13.0				1.3	1.2	0.60	0.03	0.1
I	0.93	12.02								
J	1.28	12.4								
K	0.6	13.8	0.3	0.02	0.003	2.0	2.0	(N=0.02)		

b. Mechanical Properties

	State	$RE_{0.2}$(MPa)	Rm(MPa)	A%
A	Cast & Quenched	350	1350	38
B	"	430	1230	38
C	"	430	1320	44
D	"	407	1357	54
E	"	355	673	31
F	As-Cast	345	710	27
K	Rolled & Quenched	347	841	70

c. Sensitivity to Notched Effect in Rotaring Bending Test (Kt = 3.25) (59)

Steel	Fatigue Limit Smooth Test Piece (MPa)	Notched Test Piece (MPa)	K_f	9
E	380	190	2.0	0.44
F	480	160	3.0	0.89

d. Effect of Carbide Precipitation on Fatigue Limit in Rotaring Bending Test of Steel I (61)

Thermal Treatment	Fatigue Limit (MPa)
TE 1050°C	369
TE 1050°C + 60H - 420°C	401
TE 1050°C + 8H - 580°C	411
TE 1050°C + 30H - 750°	364

e. Effect of Work-Hardening on Fatigue Limit (61)

Work-Hardening Rate %	Fatigue Limit (MPa) Rotating Bending Test (Steel I)	Stress-Strain (Steel J)
0	369	245
2		294
10	417	
25		363
30	490	
50	672	

f. Crack Propagation and Failure Parameters (59)(60)

Steel	ΔK_s(Mpa)	da/dN (mm/c)	K_c(MPa)	K_{IC}(MPa)
E	6.5	10^{-4}(ΔK/26)2.94		46-54
F	9.1	10^{-4}(ΔK/24)3.79		43-45
G	5.6		86	95.6
H	5.9		96	114.0

3. Crack growth rate

The crack propagates in three stages. During the first, the cracking rate is very slow, and for an applied stress $\Delta\sigma$, a size value "a_s" can be set, for an acceptable defect which does not cause fracture :

$$\Delta Ks = \alpha \ \Delta\sigma \sqrt{\pi a_s}$$

ΔKs = stress intensity factor
where Δs = stress, a_s = defect size

Cracking is transgranular and does not show any striations.

When *ΔK* increases the second stage is reached, which covers an extended range of variation of *ΔK*, and where the cracking rate da/dN obeys the Paris equation (figure 61) :

$$da/dN = c(\Delta K)^m$$

where c = constant
ΔK = stress intensity factor,
m = exponent

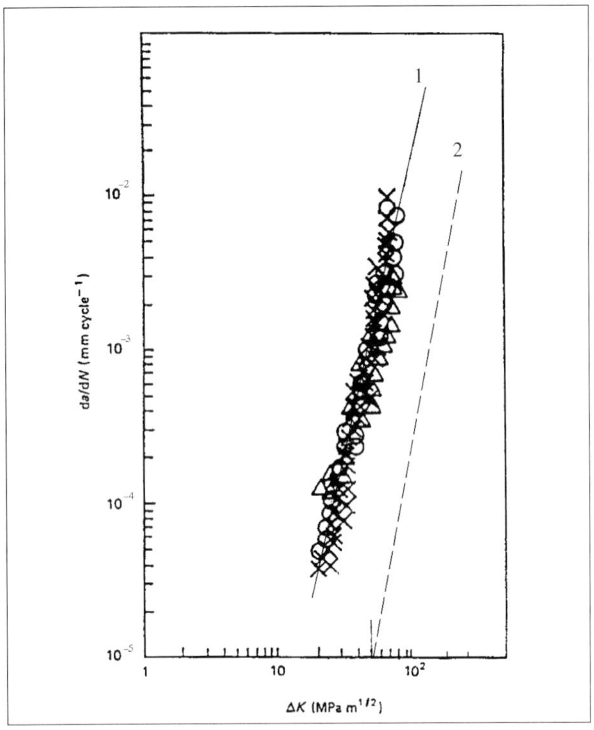

Figure 61 : Crack Growth Rate of Manganese Steels (57)(60)
1) Cast Manganese Hadfield Steel
2) Nonmagne 30

The crack surface takes on a striated aspect from a ΔK value of the order of 15 $MPa.m^{1/2}$. The pitch of the striae is independent of ΔK, but it increases rapidly in the presence of inclusions.

The crack propagation becomes more rapid, the pitch of the striae increases, and the inclusions contribute to the formation of pits when the third stage is reached, and finally a terminal fracture intervenes at a stress intensity factor of K_c, close to K_{Ic}.

4. Effect of metallurgical factors

There is little experimental data available on fatigue strength which makes it difficult to illustrate the effect of metallurgical factors.
The oligocyclic fatigue curves of four cast steels, showing differences in chemical composition, stress-strain characteristics, and grain size (from a few microns to 1 mm.) are practically the same and differ little from the curve of a rolled steel (figure 62, table 9). On the other hand, the oligocyclic fatigue curve of a fifth cast steel has a lower slope, and its mechanical characteristics do not give a satisfactory explanation for this.

In the Paris domain for the afore-mentioned steels, the cracking parameter ΔKs, and the crack propagation rate, are of similar values, only the terminal fracture parameters Kc and K_{IC} differ. The latter are in fact strongly influenced by embrittlement phenomena, although the fatigue limit seems to remain unaltered (table 9f).

Work-hardening seems to have the same effect on the fatigue limit of austenitic manganese steels as on austenitic nickel-chromium steels, the fatigue limit increasing in proportion when the work-hardening rate increases (table 9e).

The application of overstress, or repeated overstress during fatigue, produces a work-hardening at the crack tip, which slows its propagation. The length of time of propagation may increase in a ratio of 50 to 6000%, and generally a single overstress is more effective than repeated ones (60).

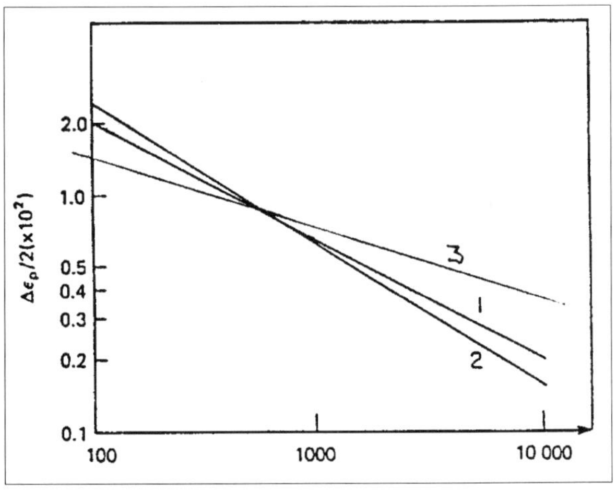

Figure 62 : Coffin-Manson Plots of Strain-Life Curves (57)
1 Cast Material, 2 Wrought Material, 3 Cast Material

IV. Embrittlement Mechanisms

The embrittlement of austenitic manganese steel is a combination of interdendritic, intergranular and at times transgranular features. The embrittling factors have only a minor effect on the shape of the tensile curve, the yield stress and the work-hardening coefficient retaining practically the same value, but fracture of the product occurs at a tensile strength and elongation values which are the more low as the embrittlement is great. The decrease of the tensile strength and of fracture elongation is accompanied by a change of aspect of the fractured surface, altering progressively from a ductile mode to a brittle mode.

Under the best conditions, ductile fracture of Hadfield steel is produced with a tensile strength of the order of 100 to 110 dN with about 70% elongation, and brittle fracture with a tensile strength of the order of 50 to 60 dN, at a few percent elongation. The elongations and the tensile strength are not independent of each other, and tensile tests prove to be a good criterion for indicating the embrittling state. The results of 400 tensile tests, carried out during the testing of industrial cast products (63) where the embrittling factors are varied (intergranular segregation of phosphorus, oxides, nitrides and carbides), led to the following relation being worked out between the elongation (ε) and the tensile strength (σ) :

$$\varepsilon\% = 1.352\,\sigma - 73.7 \quad (\sigma \text{ in dN})$$
(correlation coefficient = 0.72)

which agrees with the results of other researchers (figure 63).The observed dispersion is explained in large part by the effect of grain size and the presence of micro-shrinkages.

a) Effect of test temperature

The transition from a ductile fracture mode to a brittle one under the influence of a decrease in test temperature is progressive, and simultaneously, the yield point and tensile strength increase. In consequence the relation between the tensile strength and the elongation is different from the one obtained at room temperature. According to the results of C. Doepler (66) the following equation is obtained :

$$\varepsilon\% = 3.952\,\sigma - 364$$
Where σ is measured in dN.

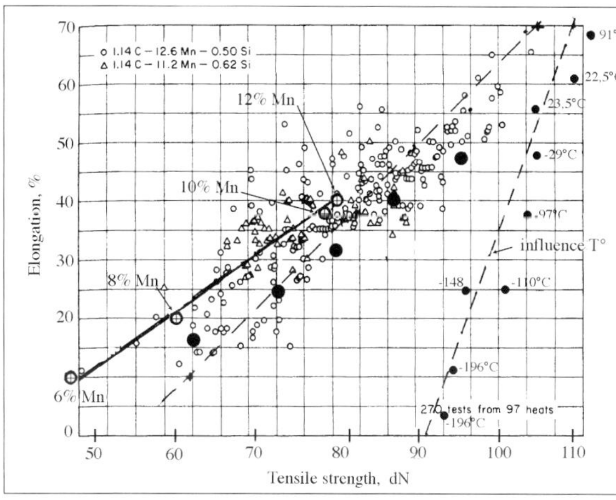

Figure 63 : Relationship between Tensile Strength and Fracture Elongations - Effect of temperature Embritting Factors . and Austenite Instability.
○ 1.14% C - 12.6% Mn - 0.5% Si - 270 tests from 97 heats (65)
△ 1.14% C - 11.2% Mn - 0.62% Si
● Hadfield Steel (64) ⊙ (75) (Effect of Mn Content)
+-- (63) (66) Effect T° 1.40% C - 12.11% Mn - 0.060% P Hot-rolled Steel (ϕ 12 mm)

b)Effect of oxides and nitrides

Silicate and oxide inclusions are particularly dangerous when they are in the form of micro- or macro-inclusions in clusters at grain boundaries. Slow-bending tests have revealed that these inclusions permit crack initiation in the surface under tensile strain, after deformation and ranging, according to their size, from 5 to 25%.
Large isolated inclusions do not initiate cracking at deformation rates less than 25-30%, unless they reach the grain boundaries, in which case the cracks are initiated at lower deformation rates of 15 to 20%.
Small spherical inclusions, regularly spread out, are less dangerous, cracking being initiated, after a large

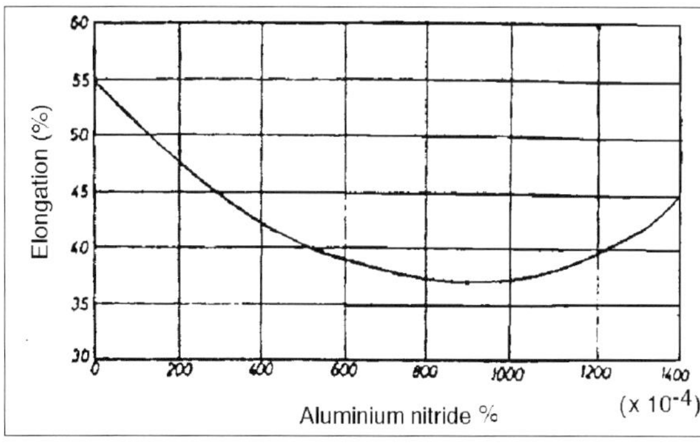

Figure 64 : Influence of Aluminium Nitride % on Fracture Elongations of Hadfield Steel (68)

deformation, only from inclusions in the triple grain boundaries (67). Aluminium nitride is precipitated in the form of fine particles in the grain and at the grain boundaries. Fracture elongation falls from 55 to 37% when the aluminium nitride content increases from 0 to 0.090% (figure 64).

c) Effect of phosphorus

Phosphorus contributes to interdendritic brittleness. An increase in phosphorus content within the solubility limit in austenite, has as a primary effect the displacement of the toughness-temperature curve, and the change to a ductile-brittle aspect of the crack, towards higher temperatures (figures 65-66). The transition temperature T50 (50% brittle - 50% ductile) of water-quenched Hadfield steel at 1080°C increases from -70°C to -45°C when the phosphorus content changes from 0.010 to 0.030%.

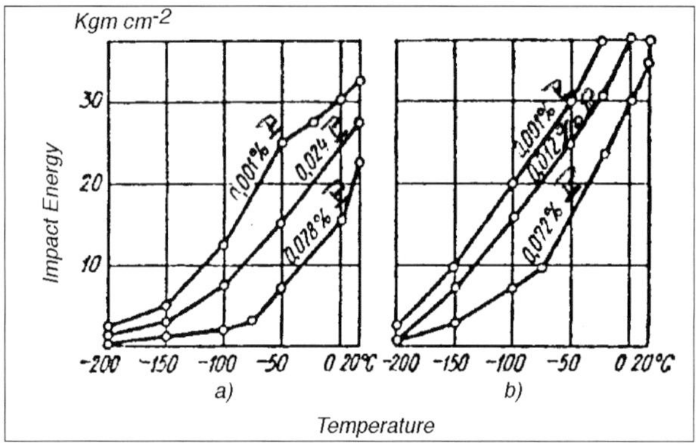

Figure 65 : Influence of Phosphorus Content on Impact Strength-Temperature Curve of Hadfield Stell
a (G13L) Cast Steel b (G13K) Forged Steel

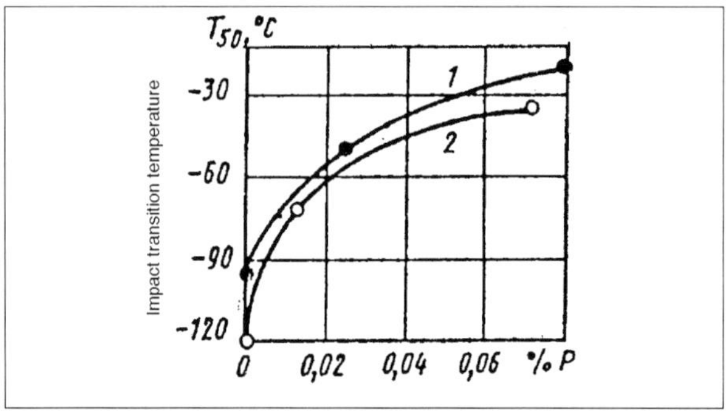

Figure 66 : Influence of Phosphorus Content on Transition Temperature T50 (appearance of the crack area : 50% ductible - 50% brittle) (55)
(G13L) 1. Steel a. cast (G13K) 2. steel b. forged
steel a - C = 1.12%, Mn = 12.5%, cast and water-quenched at 1080°C, G = 4-5
steel b - C = 1.20%, Mn = 13.0%, forged and Water-quenched at 1080°C, G = 6-7

The chemical composition of the microstructure becomes more homogeneous after reduction of the cross section by hot forging or rolling, and the effect of phosphorus on toughness is neutralized (figure 67). The question arises, how far are the mechanical properties affected in the long and short transverse directions ? At room temperature ductility decreases rapidly when the phosphide eutectic appears in the microstructure, and falls to a very low level when a continuous phosphide solution is formed (figure 68).

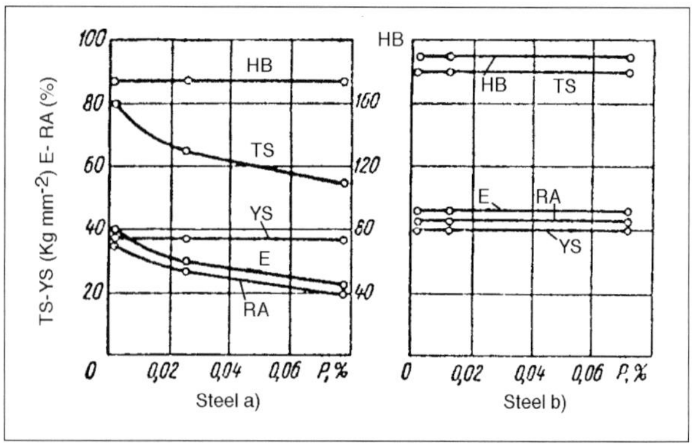

Figure 67 : Influence of Phosphorus Content on Hardness and Strength Properties (55)
G13L steel a : C = 1.12%, Mn = 12.5%, cast and water-quenched at 1080°C, G = 4-5
G13K steel b : C = 1.20%, Mn = 13.0%, forged and Water-quenched at 1080°C, G = 6-7)

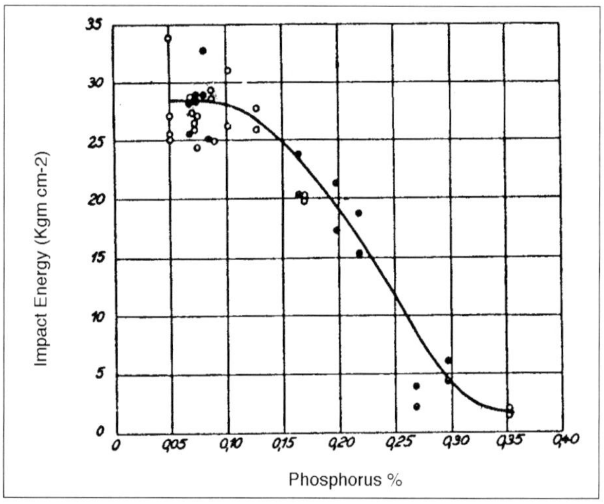

Figure 68 : Effect of Phosphorus Content on Impact Strength at Room Temperature (presence of phosphide eutectics at about P > 0.12%) (69)

d) Effect of carbides

The deflection, initiating a crack at the notch tip during slow-bending, the critical deflection ensuring its propagation, and the value of the stress intensity coefficient K, all decrease according to a logarithmic rule, when the amount of intergranular carbide, platelet carbides, and the percentage of pearlite increase (figure 69).

Intergranular carbides have a greater effect on crack initiation at the notch tip than do carbide platelets or pearlite. For example, with a deflection of 0.182 mm, cracking is initiated in the presence of 0.1% intergranular carbides, as against 1% carbide platelets or pearlite (figure 69a).

The critical deflection for propagation does not change at a carbide platelets concentration below 0.4%, and it subsequently decreases rapidly, when the percentage of carbides increases, to reach, at around 5% by volume of carbides, the deformation level corresponding to intergranular carbides (figure 69b).
The stress intensity coefficient decreases according to the logarithm of carbide or pearlite percentage, and is not affected by the kind of phases present (figure 69c).

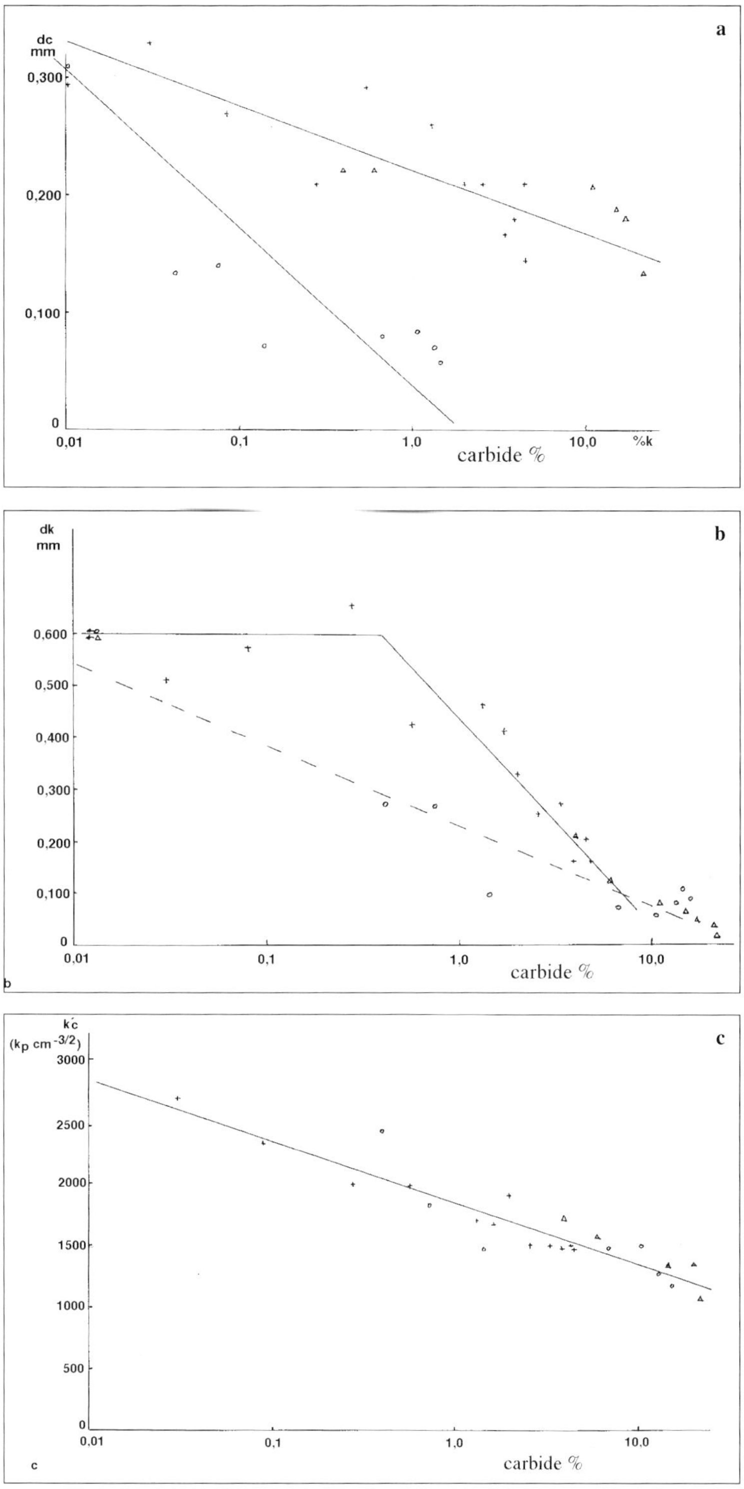

Figure 69 : Influence of Carbon % on Cracking Initiation and Propagation, Critical Deflection of Frature (70)
a) Relationship between deflection *dc* starting cracking at noch bottom and carbides %
b) Relationship between critical deflection *dk* causing fracture and carbides%
c) Relationship between stress intensity coefficient *kc* and carbides %
+ platelet carbides ❍ intergranular carbides △ pearlite + carbides + martensite

The embrittling effect of intergranular carbides does not depend solely on their volume percentage but also on the grain size. In any case embrittlement is all the more severe, all things considered, when the grain is larger. Figure 70 is an illustration of the effect of intergranular carbide percentage in relation to grain size, on elongations and toughness. For a carbide percentage of 1.8% and a grain size of 0.017 mm, elongations reach 45% and toughness 48 J, compared to the practically zero equivalent values for a grain size of 0.835 mm.

There are several sources for the carbides in the microstructure, such as an incomplete dissolution of pre-existing carbides during solution treatment, precipitation during cooling, or in re-heating. The cause is attributed to either incorrect solution treatment conditions, or a cooling rate slower than the critical quenching rate, or to an inadequate heating cycle, taking into account the kinetics of carbide precipitation in isothermal conditions.

The effect of solution treatment conditions (table 10) is shown by a series of tests carried out on crusher liners 76 mm thick, made of steel with 12% Mn, 1% Mo, and 1.19 - 1.29 - 1.44% carbon. The microstructure of the three steels, at the delivery stage, contains carbides in continuous solution at the grain boundaries. Micrographic etching shows that the dendritic structure, elongations, impact strength and toughness all have very low values. Following a solution treatment of two hours at 1090°C, the three steels still contained some isolated intergranular carbides, any dendritic structure was no longer visible under micrographic etching, and toughness was considerably improved.

The embrittlement which developed during cooling by quenching, depends directly on the kinetics of the austenite transformation in continuous cooling, no embrittlement occuring beneath the critical section of intergranular carbide precipitation. Unfortunately, not many CCT curves are available, and they are not representative of the transformation of large section cast pieces during quenching. Industrial experience has demonstrated the embrittling effect of carbon, silicon, chromium and vanadium.

The study of embrittlement, after tempering in isothermal conditions, shows the effect of chemical composition on embrittlement; as can be seen during re-heating or in the heat-affected welding zone. Carbon, chromium, vanadium and titanium, as well as nickel, which all accelerate carbide precipitation, have a negative effect on toughness. On the other hand, molybdenum, which inhibits carbide precipitation and pearlite transformation, has a positive effect. Manganese contents of between 11 and 14% do not have any significant influence (figures 71-72, table 11).

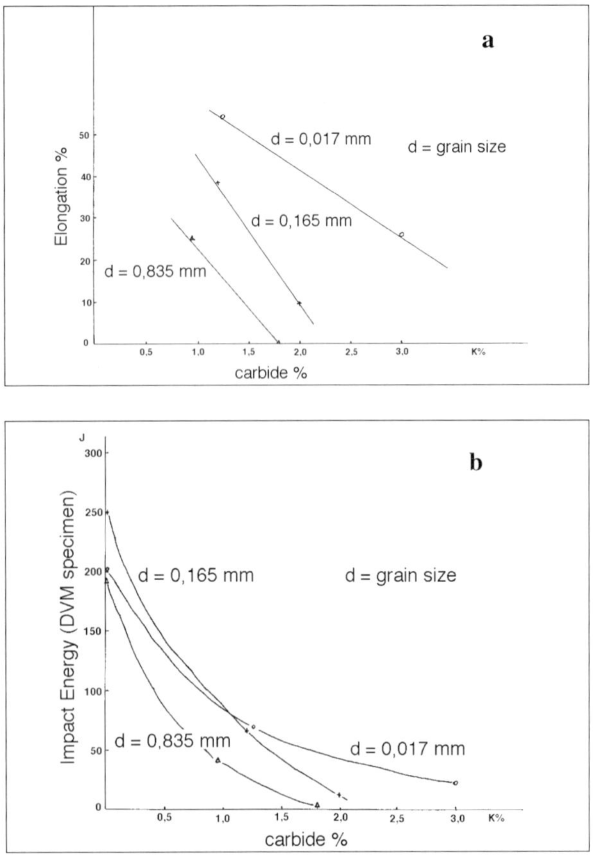

Figure 70 : Effect of Carbides % and Grain Size on Fracture Elongation and Energy (71)
a) Effect of Carbides % on Elongations for Different Grain Size Values
b) Effect of Carbides % on Fractrue Energy (Charpy DVM notched test piece for different grain size values)

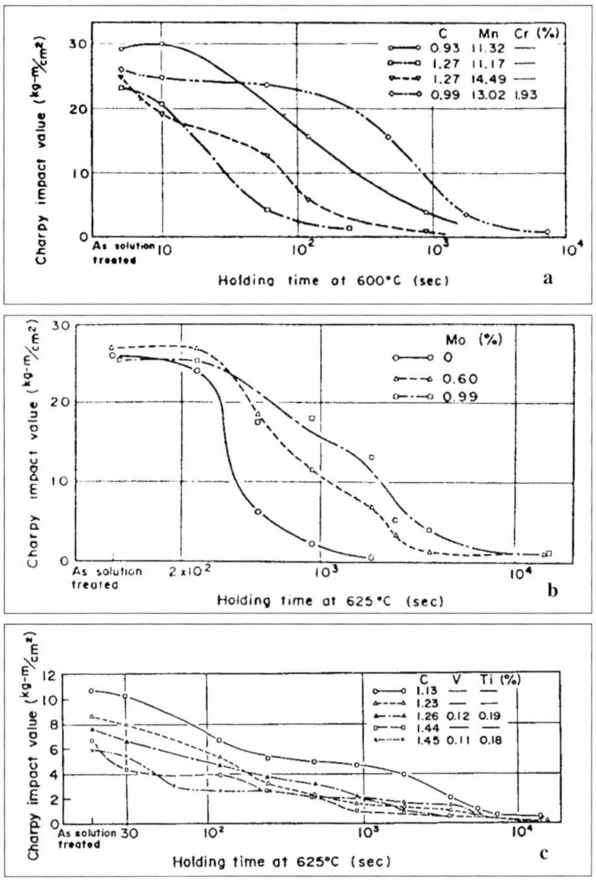

Figure 71 : Effect of Chemical Composition on Embrittlement Susceptibility (19)
a) Variation of impact value with the isothermal holding time in 13Mn cast steels (effect of C and Cr)
b) Variation of impact value with the isothermal holding time in 13Mn-2Cr cast steels (effect of Mo)
c) Variation of impact value with the isothermal holding time in 12Mn-12Mo cast steels (effect of C. V. Ti)

								Solution Treatment (2H--1090°C) & Cooled Simulating Water Quenched 76 mm Sections		Solution Treatment (2H 1090°C) & Water Quenched as 13 mm Sections	
C %	0.2% Offset Yield Strength (N/mm²)	Ultimate Strength (N/mm²)	Elon-gation,%	Hardness, HB500	Absorbed Energy V-Notch	Impact (J) EDM Notch	Fracture Toughness, K_{Id}, $MN/m^{3/2}$	Absorbed Impact Energy* (J)	Dynamic Fracture Toughness, K_{Id} $MN/m^{3/2}$	Absorbed Impact Energy* (J)	Dynamic Fracture Toughness, K_{Id} $MN/m^{3/2}$
1.19	616	636	2,25	175	14	11	32	168	88	137	85
					58			95	88	98	92
								83	88		95
									76		
1 29	371	398	2,5	170	81	45	25	60	91	141	84
	354	433	8,0		54		51	94	80	145	97
								92	88		97
								83	81		
1 44	402	470	6.5	183	23	11	54	80	68	90	77
	415	474	6.5		14	14	56	62	65	152	
								28	71		
								100	66		

Table 10 : Tensile Properties - Impact and Fracture Toughness Results for 12Mn - 1Mo Castings (72)
* Charpy specimens were EDM-notched

Composition wt%						As Quenched			Reheated 48 hr at 425 C		
C	Mn	Si	Ni	Mo	Grain Size*	Ultimate Tensile Strength kg/mm²	Elonga-tion %	Merit N°	Ultimate Tensile Strength kg/mm²	Elonga-tion %	Merit N°
1.31	12.9	0.33			CG	78	35.5	2730	61	1.0	61
					FG	94.5	42.5	4010	67.5	1.0	67.5
1.20	13.0	0.50			CG	77	49.0	3770	69	3.5	242
					FG	91	48.0	4360	70	2.0	140
1.03	12.9	0.52			CG	76	39.5	3030	66	11.0	726
					FG	92	46.0	3850	77	12.5	962
1.17	12.7	0.53	3.56		CG	79	48.0	3790	62.0	3.5	217
					FG	81	48.0	3890	63.5	4.5	285
1.15	12.8	0.51		0.98	CG	91.5	48.5	5300	76	19.0	1445
					FG	97	49.0	4750	76	23.0	1745

Table 11 : Effect of a 48-hr. Reheating at 425°C on the Tensile Properties of Austenitic 13%Mn Steel (74) * CG = coarse grained, FG = fine grained

e)Effect of plastic deformation

The low mechanical strength of the γ/α_m interface is the cause of the embrittlement of metastable austenites. The embrittlement is mostly of an intergranular type, affecting the grain boundaries more than the grain : here the deformation is greater locally, whence the preferential formation of the α_m phase and the initiation of fracture, before reaching

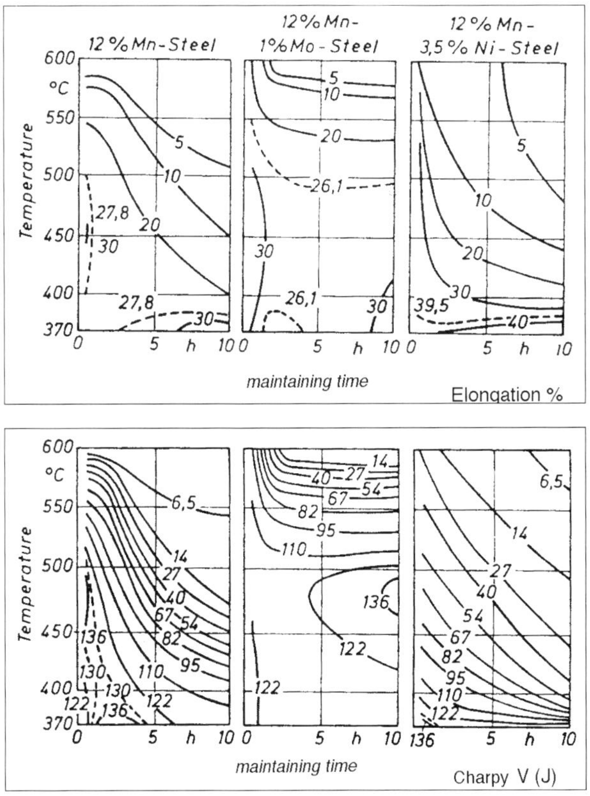

Figure 72 : Influence of Temperature and Maintening Time on Fracture Elongations and Charpy V Impact Strength of a Hadfield Steel (73)

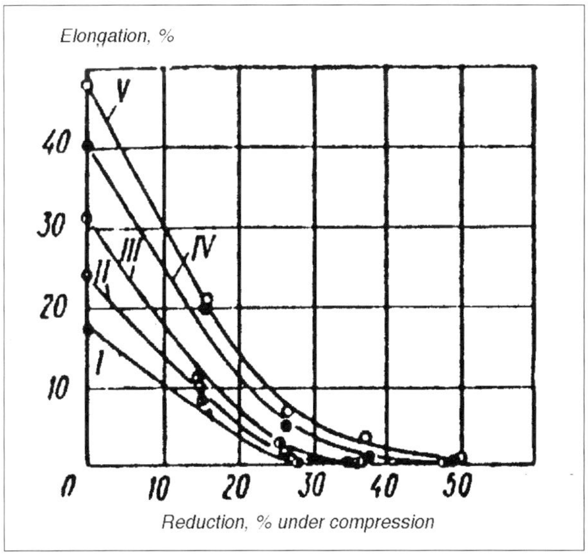

Figure 73 : Influence of Work-hardening and Embrittlement on Elongations of Hadfield Steel 110G13L (64)

	I	II	III	IV	V
UTS (kg/mm²)	63.5	72.6	79.8	90.3	98.6
A%	16.9	24/6	31.7	40.5	47.7

the grain deformation limit. The α_m phase is less harmful under compression than under tension. Intergranular cracks form under compression from a critical deformation, then the material continues to deform with these cracks until fracture, which occurs after a higher strain rate than under tension.

The elongations at fracture under tension of work-hardened Hadfield steel depend on the work-hardening rate applied and the initial state ; the less brittle and work-hardened the material is, the higher the residual ductility (figure 73). For industrial applications at low temperatures which use pre-work-hardened austenites, or which undergo deformation during service, the austenitic structure must maintain good toughness at use temperatures. An increase in manganese content and the addition of other gamma-forming elements such as nickel, copper, nitrogen or chromium, which slow down the carbon diffusion in the austenite, enable an over-rapid increase in transition temperature to be avoided, and the retention of an acceptable toughness at low temperature (34) (76) (77).

f) Toughness and critical flaw size

The critical flaw size leading to a catastrophic fracture has been calculated for the 76 mm-thick mantle of gyratory crushers, made of austenitic manganese steel, the properties of which are given in table 10. KId is the stress intensity factor measured in dynamic conditions. Using the hypothesis that maximum applied stress does not exceed the yield stress (410 Nmm^{-2}), it is found that the critical flaw size at the brittle state (KId = 33 $MNm^{-3/2}$) is 2.5 mm deep. After solution treatment of 2 hours at 1090°C (KId = 88 $MNm^{-3/2}$) the critical flaw depth changes to 18 mm (figure 74). The value of KId measured under impact bending on a fatigue-cracked Charpy test piece, can be evaluated, with some reservation, from the measurement of fracture energy on the Charpy test piece notched at a radius of 0.2 mm and 2 mm deep (machine-tooled by electro-erosion) by using the relationship [KId = 0.40 Kv + 36.7 ($MN.m^{-3/2}$)] (correlation coefficient = 0.78).

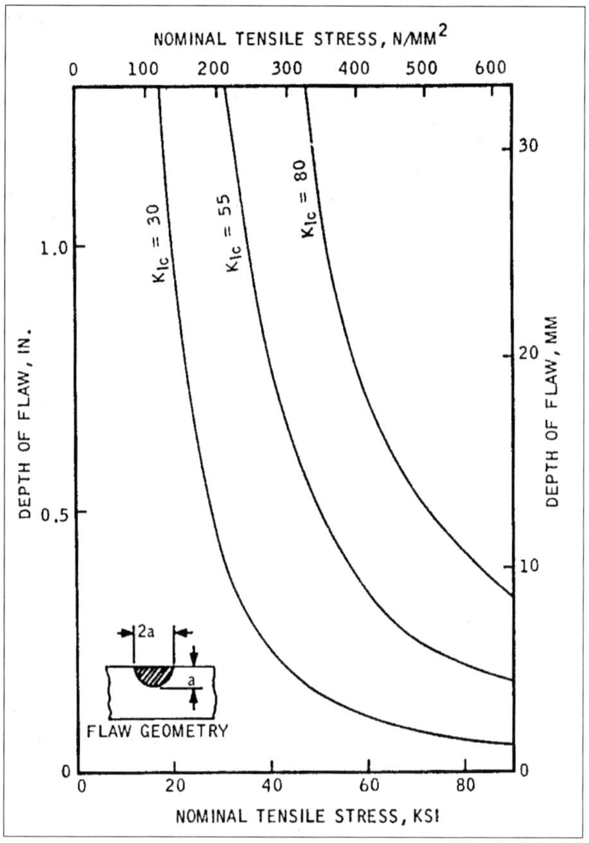

Figure 74 : Relationship between Tensile Stress and Critical Flaw Depth for Catastrophic Fracture (74)

Part 2 - Production

I. *Melting and Pouring of Austenitic Manganese steels*

II. *Role of mechanical Properties at high Temperature*

III. *Heat Treatment*

IV. *Machinability of Austenitic Manganese Steels*

I. Melting and Pouring of Austenitic Manganese Steels

The production of austenitic manganese steels requires a few precautions so as to obtain a low phosphorus content; in order to avoid the formation of oxides and nitrides and reactions with refractories, to eliminate hydrogen and to control grain size.

a) Dephosphorization

The standard two-slag method enables the phosphorus content of the metal bath to be lowered by several one-thousandth percent, but as ferromanganese contains phosphorus and cannot be added during the oxidizing period, its addition produces a substantial increase in phosphorus content.

To obtain a steel with 12% manganese and no more than 0.030% phosphorus, taking the phosphorus content of the cast to be the sum of the phosphorus in the hot metal after dephosphorization and that brought in by the ferromanganese (table 12), it can be calculated that the phosphorus content of the hot metal should not be above 0.009% if high carbon ferromanganese is used, and not above 0.029% if a low-phosphorus ferromanganese is used.

The operations of secondary melting, i.e. remelting manganese steel scraps, is carried out only under reducing conditions, consequently the phosphorus content of the tap metal corresponds more or less to the initial phosphorus content of the charge. The phosphorus content of steels produced by this method is variable, and can reach 0.070 to 0.080%.
There are two methods of lowering the phosphorus content of a metal bath containing manganese : phosphorus oxidation in the presence of lime, with oxygen or oxides, and the reaction, in reducing conditions, of phosphorus in solution with calcium.

Two alternative processes using the first method (phosphorus oxidation by oxygen or oxides) have been tested in the former Soviet Union in order to remedy the inconvenience caused by the very high phosphorus content of the high carbon ferromanganese used there (0.43-0.45% P). The one process consists of blowing a powdered mixture of lime, manganese oxide and spar in an oxygen jet (78), the other, a mixture of iron oxides, manganese and spar (79). The dephosphorization occurs by interaction with the calcium oxide of the slag.

The role of manganese oxide is to limit manganese loss, which in the presence of 20-30% MnO_3, drops from 0.8-0.9 to 0.27-0.35%; to improve the elimination of phosphorus and its transfer into the slag. Spar additions flux the slag, and improve dephosphorization. The slag should be tapped out rapidly to avoid reintroducing phosphorus into the bath.

An optimum dephosphorization rate (35-40%) was obtained by blowing a mixture of powders (5 parts CaO, 3 parts MnO_3, 2 parts CaF_2) with an oxygen jet, at 1480-1580°C. The mixture was 8-9% the weight of the charge.
Blowing such a mixture, consisting of 10-20% iron ore, 32-50% manganese oxide, 5-20% spar, of grain size below 5 mm, in 15 minutes, lowered the phosphorus content by 0.027% and the manganese content about 0.2%. Silicon loss was not above 15% in relative value, and the carbon remained unchanged.

The inconvenience of this method in the two examples given, is that a large quantity of powder is necessary. The second method relies on the reaction between calcium and phosphorus to give the phosphide Ca_3P_2, but the reaction requires large additions of calcium, several kg per ton(80).

	Mn %	C %	P % (max)	S % (max)	N % (max)
FeMn (carburized)	78.0	6.0	0.20	0.010	
Over-refined Low P	85-92	0.1-0.5	0.030	0.010	
Over-refined Low P and N	85-92	0.05-0.10	0.030	0.010	0.030
Electrolytic Mn Metal	97	0.030	0.020	0.008	0.035

Table 12 : Chemical Composition of Main Commercial Ferromanganese Alloys

The effectiveness of the treatment differs according to the phosphorus content of the metal bath, and can be strongly disturbed by extraneous reactions, the most important being the renitration of the bath in contact with the atmosphere. The dephosphorization level, at high concentrations of phosphorus ($P\ \% > 0.05$) is proportional to the quantity of calcium added. Raising the temperature increases the efficiency of the treatment, the dephosphorization rate doubles when the treatment temperature changes from 1550 to 1650°C (figure 75).

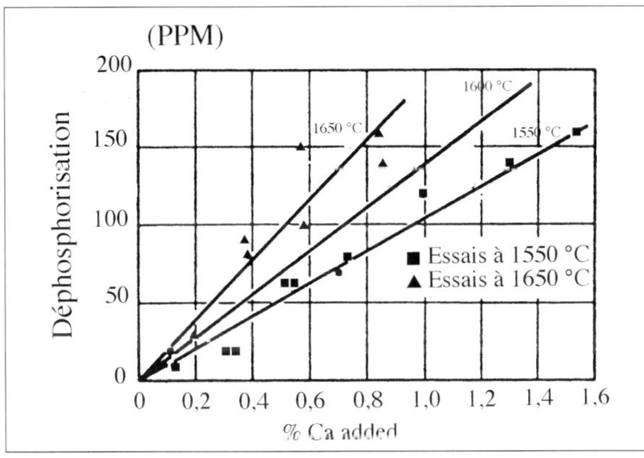

Figure 75 : Influence of Temperature on Dephosphorization of a Killed Steel with High Phosphorus Content (80)

Dephosphorization by calcium seems unchanged in the presence of CaO or of CaF_2, and is independent of the technique used, viz., powder blowing or uncoiling of calcium-cored wire. On the other hand the rate of introduction of the calcium into the liquid metal plays an essential role in the efficiency of the treatment. The dephosphorization rate of a carbon steel increases 30-50% when the rate of addition of the dephosphorizing agent changes from 0.5 to 1 kg per ton per minute.

This method allows dephosphorization down to contents of around 0.007%, and contents of N, As and Sn are greatly lowered during this treatment. It shows the inconvenience of consuming relatively large quantities of calcium, and extraneous reactions can considerably damage results. The slag finally should be oxidized after treatment to avoid the formation of the toxic gases PH_3 and AsH_3 by hydrolysis of Ca_3P_2 and As_3P_2, gathered together in the slag.

b) Oxides and nitrides

Non-metallic inclusions of manganese silicate often accompanied by oxides and aluminium nitride, have an unfavourable effect on mechanical properties (see the chapter on embrittlement of austenitic manganese steels).

The formation of inclusions of silicates, oxides, and aluminium nitride is strictly dependent on the production conditions and on the tapping. The equilibrium oxygen content of the melting bath of these steels is low, of the order of 30-40 ppm, and an aluminium addition between 0.3 to 1 kg per ton is generally used, not only to complete deoxidation, but also to avoid the formation of superficial pitting on the cast pieces (81).

Production conditions do not explain the presence of large intermediate inclusions of silicates and oxides which are sometimes observed, their morphological aspect and localization would indicate a reoxidation of the liquid metal during tapping.

The solubility of nitrogen in liquid iron is increased in the presence of manganese, and decreased in the presence of carbon. The balance favours an increase of nitrogen solubility in austenitic manganese steels at the liquid state. The nitrogen content at 1600°C, at a pressure of PN_2 of 1 At, changes from 0.043% for pure iron, to about 0.087% for a steel with 12% manganese and 1.2% carbon. The nitrogen content in industrial castings alters according to whether the manganese is introduced in the charge or in the metal bath. According to the technical literature one finds an average nitrogen content of 0.0093% in a casting bath of 4.3 tons in the electric furnace where ferromanganese was introduced into the furnace after melting the charge, and an average nitrogen content of 0.0219% in a two ton casting bath where it was introduced into the charge before melting (82).

In order to avoid aluminium nitride precipitation during heat treatment the nitrogen and aluminum contents should be in a stoichiometric ratio, and the result of their being multiplied together should not be greater than the value of the solubility product [Al][N] corresponding to the treatment temperature used. Applying this rule is difficult, the relation between the value of the solubility product [Al][N] and the temperature not being well-known : only a value of the order of $4.10^{-4}\%$ at 1050°C has been reported (68), whence the difficulty in adjusting the aluminium addition according to oxygen and nitrogen contents.

One solution which has been proposed consists of deoxidizing with aluminium (0.5 kg/t) and then fixing the nitrogen by an addition of titanium (2 kg/t). The mechanical properties should be improved, despite the generally coarse size of titanium nitride ; these results need to be confirmed (82) (83).

c) Hydrogen and decarburization

Manganese hardly changes the solubility of hydrogen in liquid iron (84) and manganese steels only show blisters at hydrogen contents above 60 cm3/100g (85). Blistering is especially observed in secondary melting operations, where hydrogen is introduced by the hydrated oxides of the charge and slag and, as the operation is conducted solely under reducing conditions, the hydrogen is not eliminated and can rise above the tolerable critical content.

Hydrogen can be eliminated by an injection of oxygen into the metal bath, and this method has also the benefit of decarburizing the metal, when the carbon content is too high. The oxidation reactions :

$$(1)\quad \underset{\text{Metal}}{[Mn]} + \underset{\text{Gas}}{\{O\}} \Leftrightarrow \underset{\text{Slag}}{(MnO)}$$

$$(2)\quad (MnO) + [C] \Leftrightarrow [Mn] + \{CO\}$$

have been studied experimentally, and the carbon-manganese equilibrium curves at different temperatures under oxidizing conditions are shown in figure 76.

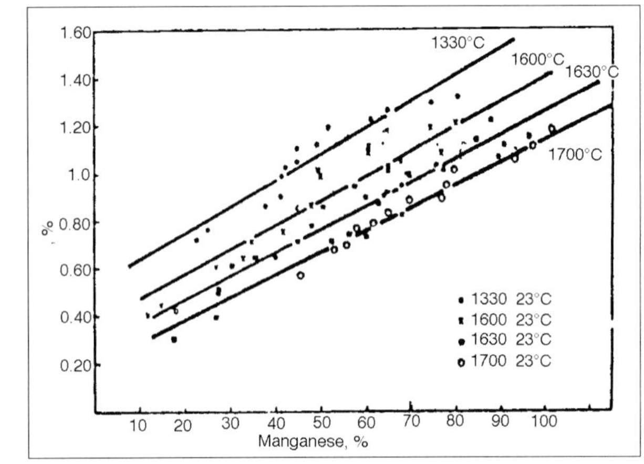

Figure 76 : Experimentally Determined C-Mn-Temperature Relationship (86)

Carbon is oxidized preferentially, for a given manganese content, when the carbon content of the bath is above the isotherm being considered ; conversely, manganese is oxidized preferentially when the carbon content is below the isotherm. Finally, the carbon and manganese oxidize simultaneously when the carbon content is on the equilibrium curve.

The study of manganese and carbon oxidation kinetics, with a view to the application of the arising results in oxygen blowing into the bath for these steels (87), shows that oxygen injection should begin at a temperature above 1600°C and that the carbon content should be maintained according to the following relation :

$$[C] > 0.017\ [Mn] + 0.20$$

Oxygen pressure, the injection lance section, and the charge weight, should be in the following relation :

$$\frac{P_{O_2}F}{Q} > 9.15 \quad 10^{-3}$$

where P_{O_2} is the oxygen pressure in kg/cm^2
F is the total lance section in cm^2
Q is the charge weight in kg.

In these conditions, deoxidation begins when the slag is saturated with manganese oxide, and after five minutes of injection, the manganese loss is 0.65% and the decarburization rate is 0.65% C/hour.

Eventually, if the operation is economical, the manganese can be recuperated by reducing the slag after the oxygen blowing.

d) Reactivity of liquid metal and oxides with refractory materials

The manganese oxide, MnO, reacts with acid oxides, giving two double oxides with silica, the rhodonite MnO SiO_2 and the tephroite 2MnO SiO_2; these oxides forming, with the silica, a ternary eutectic, which has a melting point of about 1251°C. Manganese oxide does not react with zircon (ZrO_2 SiO_2), chromite $(FeMg)O(CrAl)_2\ O_3$, olivines $(FeMg)_2\ SiO_4$, aluminium oxide or magnesium oxide

Zircon and chromite are partially reduced by the reducing elements of the melting bath and they release, respectively, silicon (from SiO_2 in zircon) and chromium.

The liquid metal oxidizes during the melting of the charge, during ladle tapping and in the mould. When the manganese content of the melting bath rises above about 5%, the mass fraction of oxidized manganese is sufficient to attack the siliceous lining of the furnaces and tapping ladles, causing crater-like surface defects on the castings, and to change the chemical composition of the surface layers. Reduction of moulding sand to chromite and the zircon coating by the reducing elements of the bath, is limited to an alteration of the chemical composition of the surface layers.

The carbon and manganese contents near the surface of a thin section of Hadfield steel cast in siliceous sand, can fall to 0.4% carbon and 6.0% manganese. The depth of the affected zone can reach 0.5-3.0 mm, before nominal contents are encountered again. On parts of Hadfield steel obtained through moulding with chromite moulds, a 0.2 mm thick surface layer enriched in chromium and containing 2% chromium at the surface has been seen, and with a zircon coating, the silicon in metal was close to the initial one.

The use of olivine sand, chromite and magnesia linings has become widespread in foundries to avoid defects and surface changes. Alumina linings of casting paddles have the double advantage of not being attacked by manganese oxide, and having a low thermal conductivity, which limits heat loss and allows operation at a lower and more regular casting temperature (88) (89) (90) (19) (91) (92).

e) Control of grain size

The average diameter of the grain of Hadfield steel in the as-cast state, according to the section of the pieces and the production conditions, is between 0.05 and 10 mm. The grain size follows a logarithmic Gaussian distribution (figure 77).

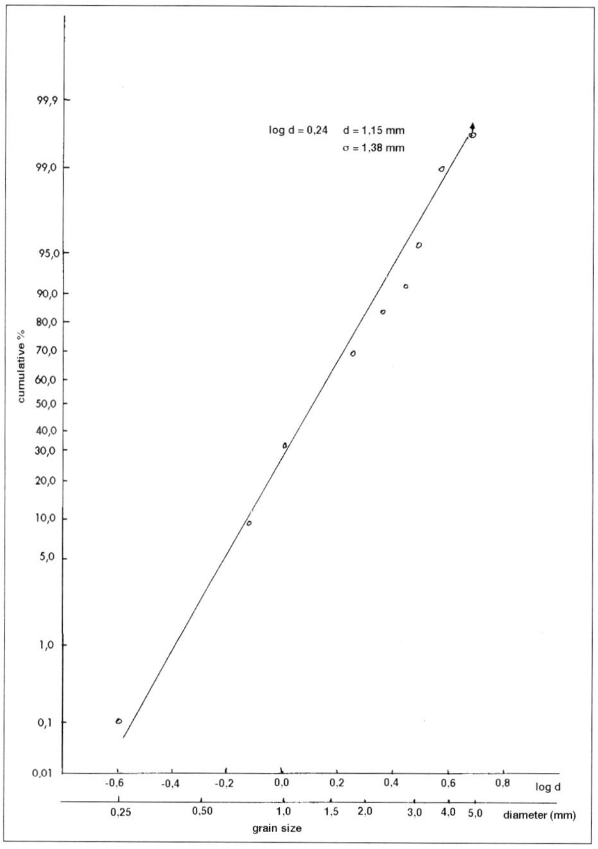

Figure 77 : Logarithmic Normal Distribution of Grain Size of Hadfield Steel (Cumulative % - log d - Grain Diameter (mm))

The casting temperature and the solidification rate are the main parameters governing the grain size. The ASTM index of the grain size measured on mouldings of 20 mm in diameter, passes from 5 to -2 when the casting temperature increases from 1500 to 1640°C (figure 78). A difference of two points can be noted in the index of grain size between a rapid solidification in a metallic mould and a slower one in dry sand (figure 79).

Increasing the moulded section has a considerable influence on the grain size, and the coefficient changes from 5 for a diameter of 45 mm, to a coefficient of -7 to -8 for a diameter of 100 mm (figure 79).

Attempts to refine the grain have been directed towards the possibility of encouraging austenite nucleation in the liquid metal by blowing in metal powders or by the addition of carbide-forming elements (Cr, V, Ti, Nb, Ce) which are capable of supplying carbide nuclei to the melting bath (93) (94) (95) (15). Blowing in powders of the FeMn alloy has been shown to be without effect ; adding carbide-forming elements enables the grain of Hadfield steel

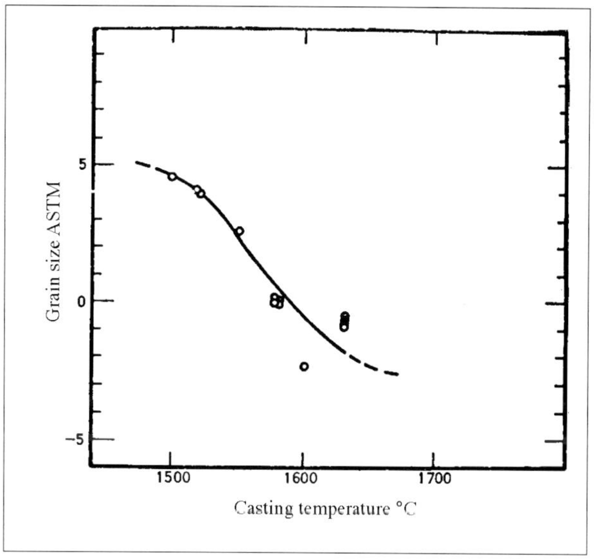

Figure 78 : Influence of Tapping Temperature on Grain Size of a Hadfield Steel (20 mm diameter section)

to be refined ; chromium at a relatively high content, vanadium at a low content (figure 80). The effect of titanium, niobium and cerium on grain refining seems similar to that of vanadium.

The size of the grain is not only governed by the parameters of solidification, the grain size increases after solidification during cooling, between the solidus temperature and a temperature more or less low depending on the chemical composition. The carbide-forming elements have a strong effect on the kinetics of grain growth (figure 81).

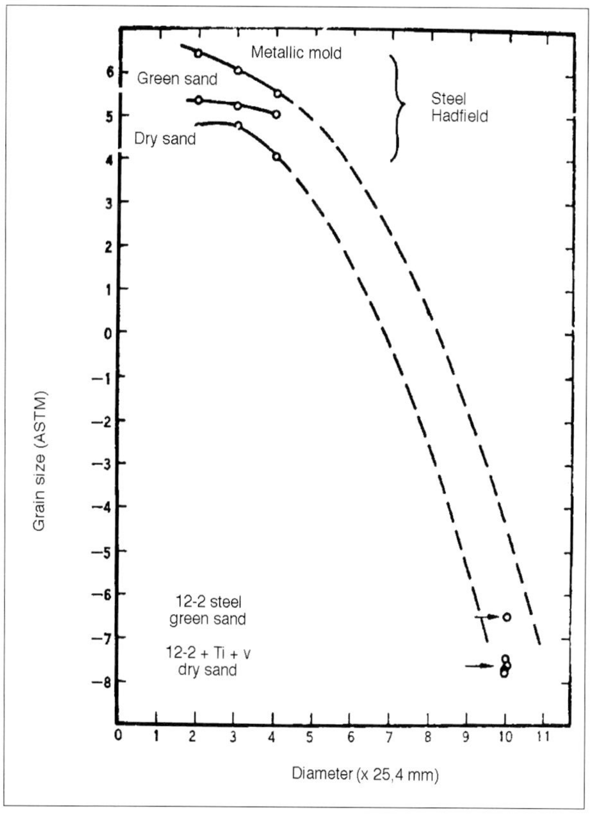

Figure 79 : Influence of section on Grain Size of a Hadfield Steel (Tapping Temperature 1500°C))

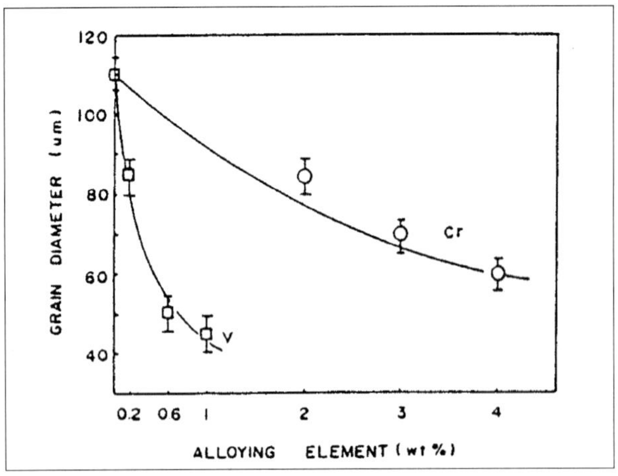

Figure 80 : Influence of Cr and V on Grain Size of Hadfield Steel (94)

As a general rule, to obtain a fine grain size, the casting temperature should be as low as possible, taking into account the need for the mould to be well-filled. The casting of thin sections offers several possibilities (in addition to the pouring temperature) for obtaining a finer grain, e.g. thermal diffusivity of the mould to increase the solidification rate, stripping and quenching at high temperature, and addition of carbide-forming elements to inhibit grain growth during solidification and cooling. Henych (93) obtained a finer grain by accelerating the cooling of Hadfield steel test pieces cast in a metallic mould, by water-quenching after stripping at 1050°C.

Industrial experience has shown that the thermal diffusivity of the mould and the addition of carbide-forming elements are insufficient for refining the grain of heavy section castings of Hadfield steel. The mathematical solidification models show that the cooling rate of the metal layer, taken between the liquidus and the solidus temperatures, follows a decreasing rule when the solidified thickness increases, and that the solidification time of this layer, proportional to the thickness at the beginning of solidification, then increases rapidly with an increase in thickness. Heat exchange is thus mainly governed by the thickness of the solidified layer, from a given size, solidification rate and the thermal gradient decrease and all conditions are satisfied for ensuring dendritic growth and the production of a coarse grain size. To obtain a grain size as small as possible in heavy castings the pouring temperature is the main parameter requiring control.

Figure 81 : Influence of Maintening High Temperatures on Grain Size of a Hadfield Steel with Different Additions of Alloying Elements (97)
1) Hadfield Steel 110G13L 2) Addition of 0.074% Ce
3) Addition of 2.30% Ni and 0.03% Ti
4) Addition of 1.11% Cr and 0.14% Ti
5) Addition of 0.052% Ti

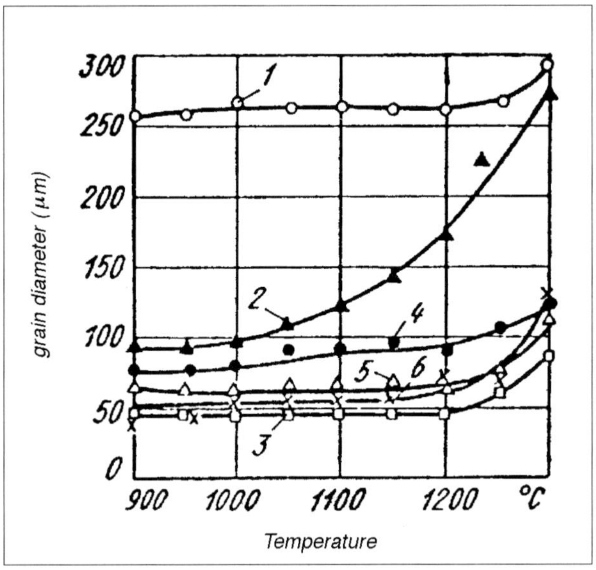

II. Role of Mechanical Properties at High Temperature

The high temperature mechanical properties of manganese austenites come into play during different stages of production : solidification, rolling, heat treatments, welding and grinding. The high value of shrinkage and expansion indices, the low heat conductivity and the effect of embrittling factors on hot ductility can all lead to production problems.

a) Ductility at High Temperatures

The temperature of manganese austenitic steels at which ductility is zero depends on the solidus temperature, eutectic formation, and liquation phenomena at low melting point. The high interdependency of carbon content and zero-ductility temperature, and the low interdependency of the latter with manganese content, are, qualitatively, in good accordance with the effect of these two elements on solidus temperature (figures 82, 84a, 84b).

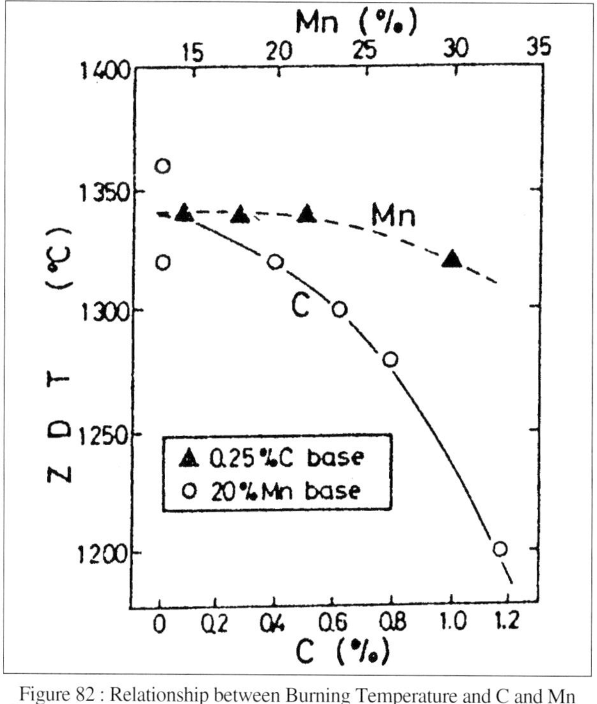

Figure 82 : Relationship between Burning Temperature and C and Mn Contents (hot stress testing carried out on test pieces taken perpendiculary from the columnar structure - deformation rate $1.10^{-3}s^{-1}$) (98)

Phosphorus lowers the zero-ductility temperature by 150°C when P content goes from 0.022 to 0.030%

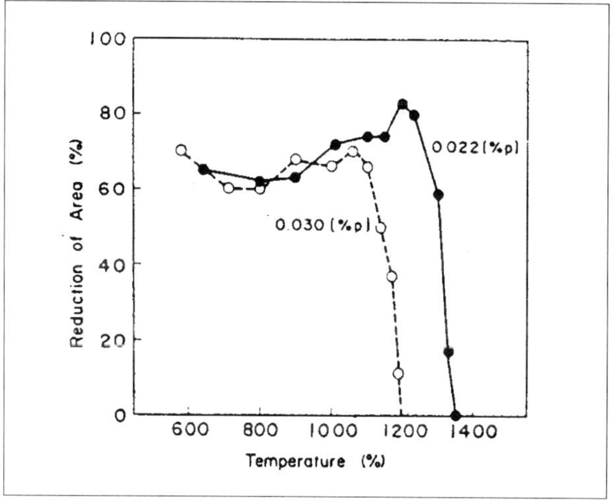

Figure 83 : Hot Deformability at Different Phosphorus Content (14)
- Tensile testing on test pieces taken perpendiculary from the columnar structure
- Deformation rate $0.3.s^{-1}$
- C% 1.05, Si% 0.56, Mn% 13.24, S% 0.008, Ni% 0.02, O 25 ppm, N 95 ppm

(figure 83). The harmfulness of this element does not depend solely on weight content, but also on the segregation rate, grain size and texture. Comparing figures 83 and 85a shows in fact that for the same phosphorus content, the zero-ductility temperature of a uniaxial structure with coarse grain size, is inferior to that of an equiaxial fine-grained structure.

Moderate contents of molybdenum, of the order of 1%, do not affect the zero-ductility temperature ; at high contents, and in association with phosphorus, molybdenum can produce liquation phenomena which lower the zero-ductility temperature by a large margin. For example, the zero-ductility temperature of a Hadfield steel with 0.035% phosphorus, is lowered by 75°C in the presence of 2.0% molybdenum (figures 85a, 85b). Increasing chromium content from 5.0 to 10.0% lowers the zero-ductility temperature by 50°C (figure 86).

Molybdenum has a tendency to reduce hot ductility (table 13), but on the other hand it has the advantage of increasing the yield strength and hot fracture resistance. The effect of silicon on weldability indicates that this element has an adverse effect on hot mechanical characteristics.

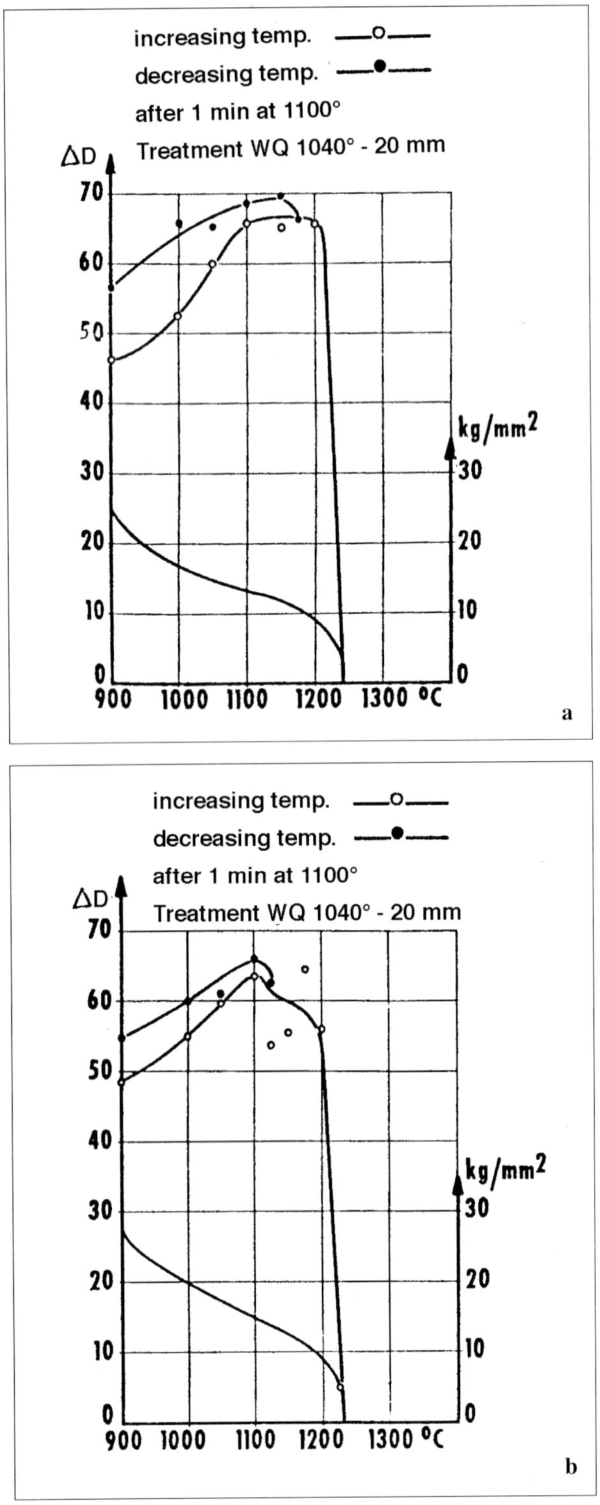

Figure 84 : Effect of Addition of Cr and Mo on Hot Deformation of a 6% Mn Austenitic Steel - Tensile Test - Rolled Steel (99)
a) C = 1.14%, Si = 0.26%, S = 0.15%, P = 0.035%, Mn = 12.7%
b) C = 1.21%, Si = 0.36%, S = 0.16%, P = 0.035%, Mn = 12.9%, Mo = 2.02%

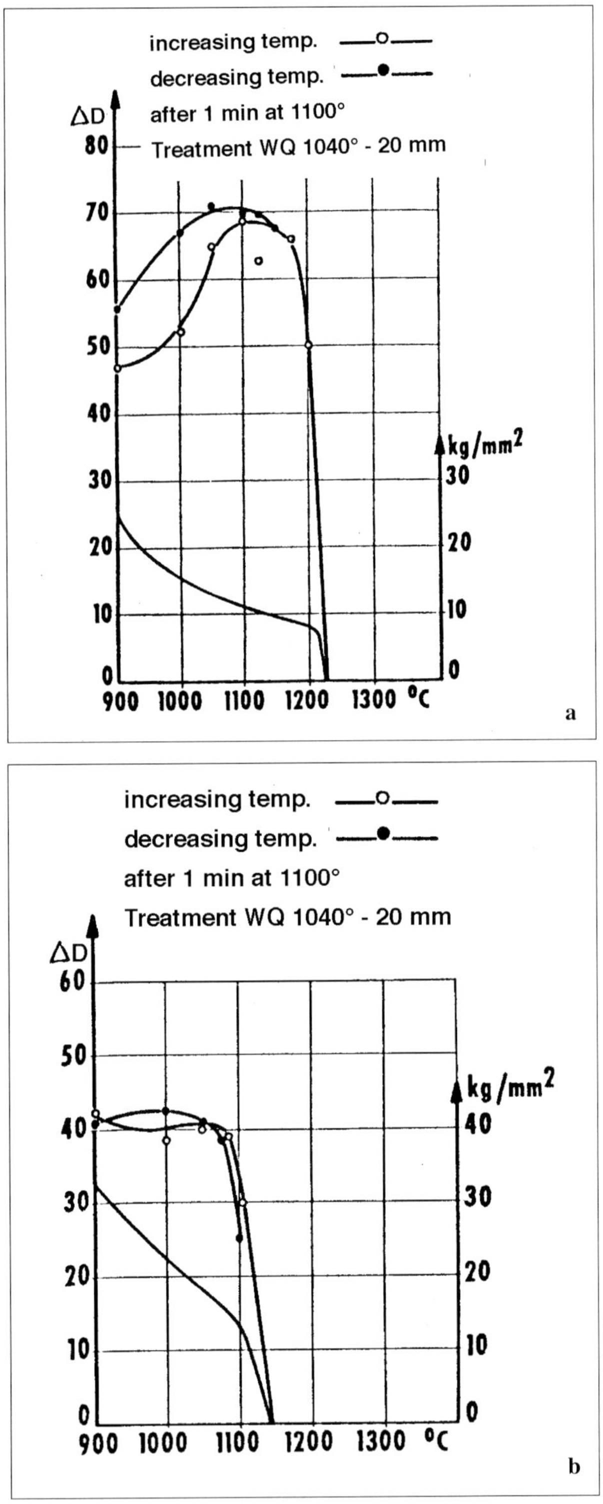

Figure 85 : Effect of Addition of 0.035% P and 2.0% Mo on Hot Deformation of a 12% Steel - Tensile Test - Rolled Steel (99)
a) C = 1.20%, Si = 2.55%, S = 0.012%, P = 0.011%, Mn = 6.5%, Mo = 0.031%
b) C = 1.18%, Si = 0.29%, S = 0.020%, P = 0.023%, Mn = 5.98%, Cr = 1.74%, Mo = 0.937%

b)Hot cracking

The hot cracking of castings occurs through the effect of the stress engendered by forces exerted between the casting and the mould, during solidification and under the effect of internal strain induced in the parts, during heating and cooling, by the thermal gradients.The study of the appearance of cracks at different stages in the production of crusher jaws of Hadfield steel (145 heats) has shown that about 33% of the cracks appear after stripping, 46% during solution treatment, and 13% after austenizing, during

	C	Si	Mn	P	S	Cr	Ni	V	N
A	0.56	0.46	17.7	0.017	0.002	5.32	0.96	0.51	0.031
D	0.58	0.55	18.0	0.019	0.009	10.84	1.00	1.90	0.027
E	0.56	0.64	18.1	0.020	0.008	10.56	1.02	0.98	0.091

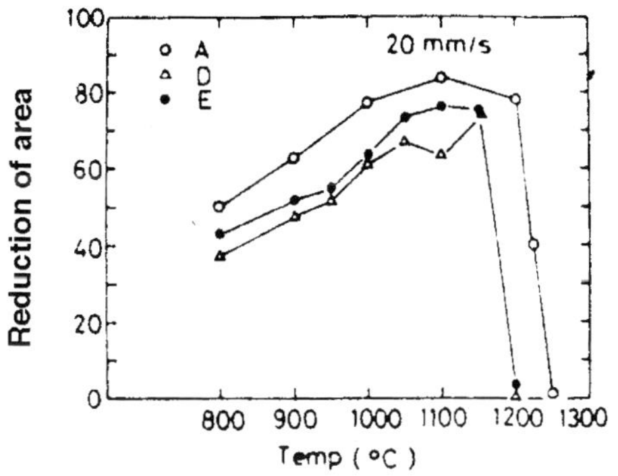

Figure 86 : Hot Tensile on Test Pieces Taken from Hot-rolled Plate - Deformation Rate 20 mm s^{-1} (100)

Steel Designation	0.2% Offset Yield Strength, MPa	Ultimate Tensile Strength, MPa	Elongation in 25 mm (1 in.), %	Reduction in Area, %
1.15C-12Mn	39.3	42.7	11	13
1.15C-12Mn-0.5Mo	46.2	48.3	12	11
1.15C-12Mn-1Mo	49.0	54.5	7	12
1.4C-12Mn	43.4	53.1	22	27
1.4C-12Mn-0.5Mo	44.8	53.8	28	42
1.4C-12Mn-1Mo	51.0	59.3	22	30
1.2C-6.5Mn	32.4	42.7	39	44
1.2C-6.5Mn-0.5Mo	31.7	41.4	29	32
1.2C-6.5Mn-1Mo	38.6	44.1	30	52

Table 13 : Tensile Properties at 980°C (1800°F) of as-Cast Manganese Steels (101)

quenching. The cracks all have the same morphology, intergranular cracking with large grain size, and show microshrinkages : they can be differentiated however by the crack opening and the oxidation state (102).

Cracking susceptibility during solidification depends on casting conditions, mould rigidity, casting temperature, and elements controlling the burning temperature. Thermal cracking is inhibited by an increase in hot yield and tensile strengths, and by a slowing of the kinetics of carbide precipitation, and of austenite transformation.

c) Weldability

Weldability is connected in a specific way to hot cracking of castings, cracking of the weld seams during solidification, and thermal cracking in the heat-affected zone. Metallurgical factors have a similar effect in these two cases.

Manganese austenites, along with other ferrous alloys, suffer from the adverse effect of carbon on weldability. Cracking in the heat-affected zone of a manganese austenite, under the test conditions which were used, occurs at a carbon content of the order of 0.8%, and the number of cracks increases rapidly with the increase in carbon content. The impact strength of the heat-affected zone, of the order of 150 J cm^{-2} at 0.8% carbon, is very low at C = 1.2% (figure 87).

Phosphorus and silicon are unfavourable for weldability. The cracking rate of the weld metal increases with an increase in the silicon and phosphorus content of the base metal or the filling metal, and when test conditions are severe (figure 88).

Carbon and phosphorus contents considerably limit the permissible heat input during welding, the critical value of the relation energy (E)/length of weld metal produced (l), for a welding rate of 30 cm mm^{-1}, is given by the relationship :

$$\frac{E}{l} = 37.5 - (54.8\ P\% + 22.0\ C\%)\ \text{KJ cm}^{-1} \quad (105)$$

An interaction between the phosphorus and the aluminium has been noted in electroslag welding, and a residual aluminium content of less than 0.06% is recommended (106).

Up to 0.050% sulphur does not affect crack sensitivity of the heat-affected zone, which is explained by the uniformity of sulphur in the microstructure, contrary to the segregation of phosphorus, and to the effect of silicon (103) (107).

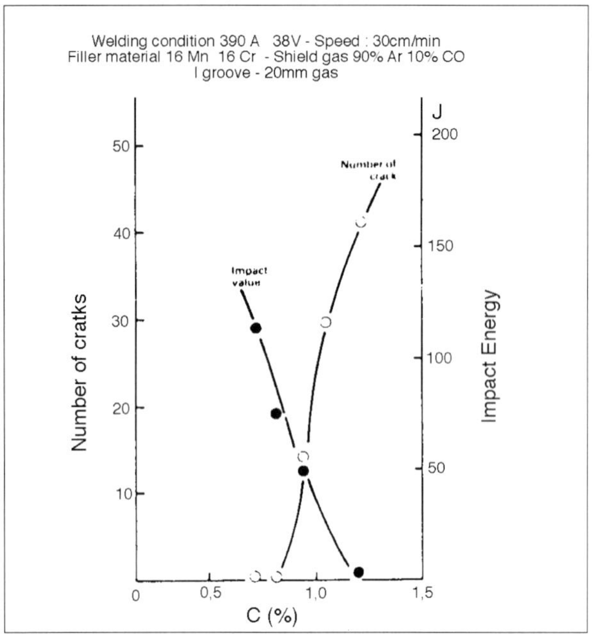

Figure 87 : Effect of C Content on the Number of Cracks and the Impact Strengh of the Heat-affected Zone for 13% Mn - 0.025% P Austenitic Steels (103)

These data demonstrate that austenites at less than 0.8% C with low phosphorus content and a moderate silicon content, are weldable. Grades with higher carbon contents can withstand welding, separation and hard-facing operations if certain precautions are taken.

Weldable manganese austenitic steels

Weldable manganese austenitic steels are particularly attractive in applications where non-magnetic material is used. Based on weldability tests, it can be hypothesized that the carbon, phosphorus and silicon contents of a weldable manganese austenite, should be less than 0.8, 0.020, and 0.5% respectively.
The mechanical properties and the magnetic permeability of the welded joints, according to standard welding methods, of several weldable rolled grades, are shown in table 14. These steels are insensitive to cracking and the mechanical properties, and the magnetic permeability of the welded metal and of the heat-affected zone, are of the same order of magnitude as that of the base metal.

The rate of fatigue crack propagation in a weldable manganese austenitic steel is identical to that of a 304 stainless steel ; the rate is lower for a same stress intensity index value DK, in the welded metal and the heat-affected zone, (sheet 50 mm thick welded by electronic beams than in the base metal (figure 89a).
The propensity to sudden propagation in brittle mode of a pre-existing crack, in a weldable manganese austenitic steel, is much less than in constructional steels killed with aluminium. The critical value at -50°C of crack opening displacement (COD) is identical for the base metal, the weld metal, and the heat-affected zone (figure 89b).

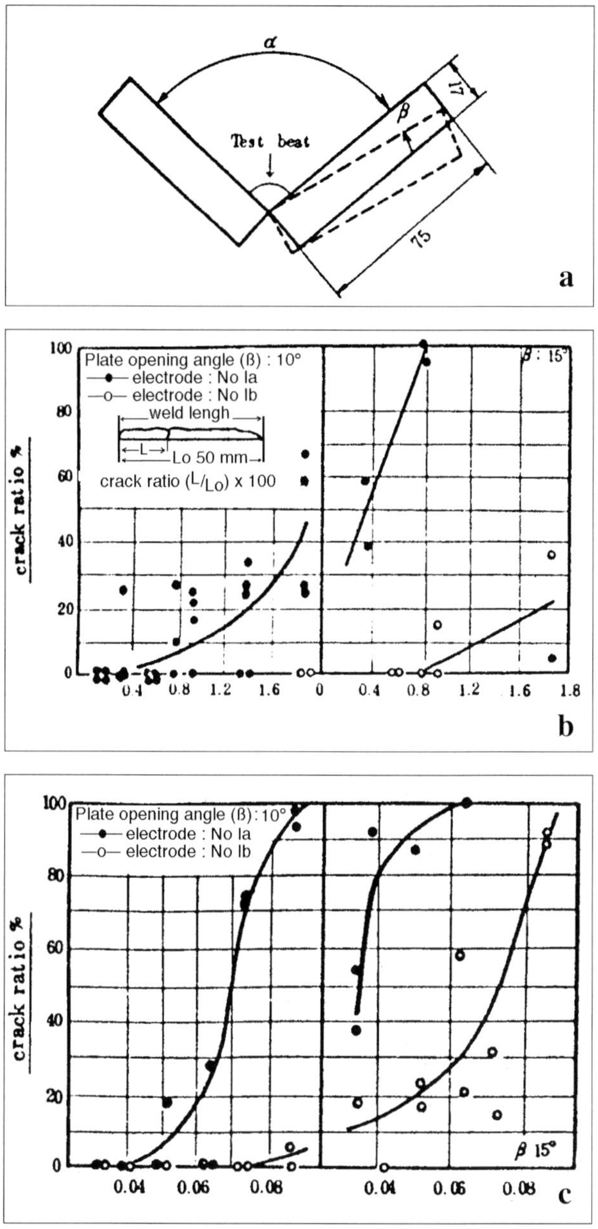

Figure 88 : Influence of Si and P on Crack Sensitivy of Weld Metal of a Hadfield Steel (104)

a) Welding conditions, and test method
Welding angle : α = 100°C, Bending angle : β = 10 and 15°,
Bending speed : 0.46 - 0.69° s^{-1}
Electrodes : steel 16 Mn - 16Cr - 2.3 Ni 1a 0.09C% - 0.014 Si% / 1b 0.15C% - 0.51 Si%
Intensity : 104 A, Welding speed : 180mm.mm^{-1}, Welding time : 21.7s

b) Influence of Si content in base metal %, in Hadfield steel, and of Si and P content of electrodes on crack sensitivy of the weld metal

c) Influence of P % content of base metal, in Hadfield steel, and of Si and P content of eclectrodes on crack sensitivy of the weld metal

Post-welding heat treatment enables the internal stresses to be relieved, e.g. by tempering for 4 hrs. at 600-650°C the residual stress level is lowered by about 40%, and a treatment of 4 hrs. at 700°C is needed to reduce the residual stress to less than 100 MPa; the stress-relief treatment lowers the toughness (as measured by a notch charpy test at 0°C) of the

Welding Process	Base Metal	Designa -tion	Filler Material	Tensile Properties (MPa)		Impact Value (J) $_vE_0$			Permea bility of Weld Metal
				0.2% PS	TS	Weld Metal	Bond		
Manual Arc Welding	0.7C-15Mn-1Ni	RM711	0.6C-15Mn-4Ni	345.0	778.1	125.4	120.7	114.5	1.002
	0.7C-15Mn-1Ni-0.6V	RM720	0.6C-15Mn-4Ni	424.6	793.4	122.8	118.6	113.7	1.002
	0.6C-14Mn-2Cr-2Ni		0.2C-16Mn-16Cr-2Ni	568.4	788.9	105.8	130.3	169.5	≤1.005
	0.7C-16Mn-2Cr	NMII	0.2C-16Mn-16Cr-2Ni	--	741.9	136.2	78.4	65.7	1.003
	0.5C-18Mn-5Cr	ASTM A289B	0.2C-16Mn-16Cr-2Ni	--	758.1	135.2	121.5	177.4	1.005
	0.6C-24Mn-6Cr	RM801F W	0.7C-15Mn-4Ni	434.4	814.9	125.4	159.7	183.3	1.002
MIG Welding	0.7C-15Mn-1Ni-0.6V		0.6C-24Mn-6Cr	--	863.4	118.6	115.6	120.0	1.002
	0.6C-14Mn-2Cr-2Ni		0.4C-14Mn-9Ni-2.5Cr	464.8	717.8	120.5	135.2	153.9	1.003
	0.6C-24Mn-6Cr		0.6C-24Mn-6Cr	--	869.3	12.05	102.9	118.6	1.002
Electron Beam Welding	0.7C-15Mn-1Ni-0.6V		--	426.6	897.3	118.6	--	117.6	1.002
	0.6C-14Mn-2Cr-2Ni		--	388.3	854.3	200.9	205.8	206.8	1.004
	0.7C-16Mn-2Cr		--	--	837.9	78.4	76.4	84.3	1.003
	0.6C-24Mn-6Cr		--	354.8	873.2	151.9	--	186.2	<1.003

Table 14 : Mechanical Properties and Magnetic Permeability of Welding Joints made by Different Welding Processes (108)

base metal, the welded metal and of the heat-affected zone. The toughness level after treatment is acceptable (110).

Not-easily-weldable manganese austenitic steels

Welding operations are carried out on Hadfield steels in repairing, hard facing and assembling. The high carbon content of these steels considerably lowers their weldability, as shown by weldability tests.To limit the difficulties, the phosphorus content should be kept as low as possible, and the presence of alloying elements which change the thermal stability of the austenite should be avoided. An addition of 0.5 to 1.0% molybdenum should be beneficial through its favourable effect on hot mechanical properties and the thermal stability of the austenite.

Welding procedures recommend no pre-heating, welding in cold conditions at less than 20 KJ cm^{-1}; to avoid any overheating by limiting, if need be by spraying, the temperatures between weld deposits to around 100°C. The filler metals used are mentioned in table 15.

d) Thermo-mechanical transformations

The columnar structure and the presence of microshrinkage in Hadfield steel ingots, require precautions to be taken, during the first rolling passes, to break the primary structure and reweld the pits, in order to give the material sufficient mechanical resistance against subsequent rolling operations. For example, the roughing, of continuous casting slabs 200 mm in width and of entirely columnar structure, until a reduction rate of the order of 50% is reached. Then, the slabs are water-quenched and examined : fine, superficial transversal cracks and edge cracks can be seen, both hardly visible in slabs in the as-cast state, and internal cracks situated at depths of between 10 and 40 mm from the surface. Internal cracks approach the surface sufficiently during rolling to intersect the scale layer in formation.

All these cracks propagate in the interdendritic areas, and their formation is due to the effect of the dendritic segregation of phosphorus on hot deformation, in particular on the decrease in burning temperature. A phosphorus content of the order of 0.022-0.025% prevents the formation of internal cracks, and reduces fire loss (14).

The dynamic recrystallization rate of Hadfield steel is rapid in the γ zone, above 825°C. A cross-section reduction ratio of the order of 3.0, is sufficient to stabilize grain size, and obtain optimal mechanical properties in the quenched state (14).

The dynamic recrystallization rate decreases considerably in the γ + carbides field. After deformation of 30 to 50% at 800°C, a structure is obtained comprising polygonized coarse primary grains with a high dislocation density in the sub-boundaries, surrounded by recrystallized grains. The primary grains disappear when the deformation rate exceeds 75%, and in their place are found recrystallized grains and globular carbide precipitates in the grain boundaries.

A decrease in the thermo-mechanical treatment temperature slows dynamic recrystallization and increases carbide precipitation. Below 750°C, the recrystallized volume fraction is small, even at high deformations, and the deformed primary grains present cellular dislocation arrangements with

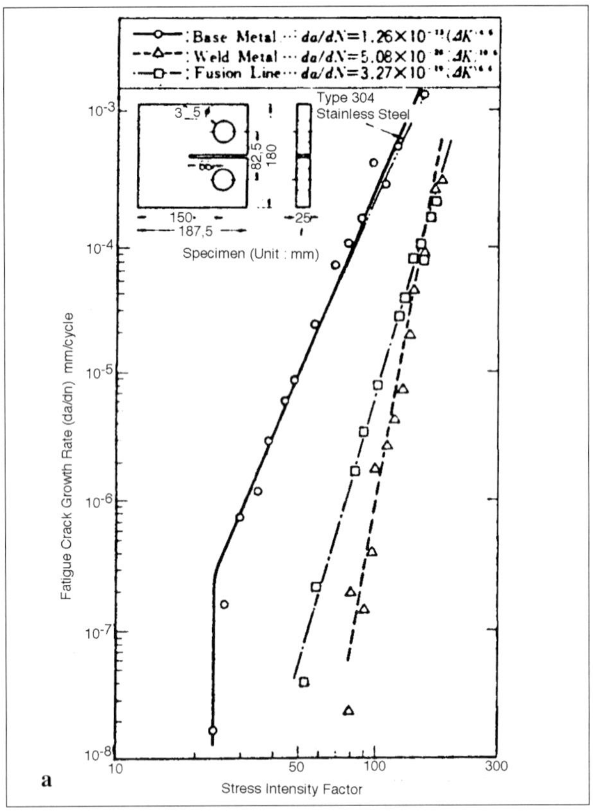

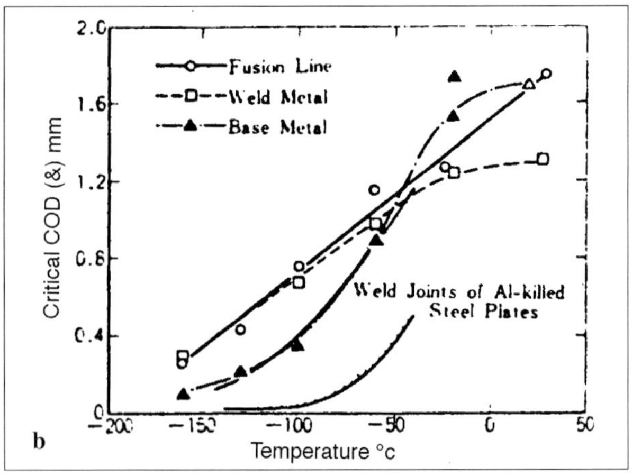

Figure 89 : Critical Opening Displacement and Propagation Rate of a Fatigue Crack of Non-magnetic 30 Steel (109)

a) Propagation rate of a fatigue crack in the base metal, the weld metal, and the heat-affected zone, of non-magnetic steel 30, and in stainless steel 304
C% = 0.55-0.65, Mn% = 13-15, Si% < 0.50, P% ≤0.030, S% ≤0.020, Ni% = ≤ 2.20, Cr% ≤ 2.20, N% ≤ 0.050

b) Effect of test temperature on critical COD - non-magnetic 30 steel

numerous carbide precipitates in the grain boundaries (112) (113) (114).

Thermo-mechanical treatments, compared to a rolled and quenched treatment, enable the production of a greater hardness and yield strength, to the detriment, of course, of toughness. Nevertheless, a hot Thermo Mechanical Treatment for the Hadfield Steel molybdenum grade increases the yield strength considerably more than carbide precipitating does. The addition of titanium or vanadium, eventually in combination with other alloying elements, modifies the dynamic recrystallization, carbide precipitation and amplifies the effect of thermo-mechanical treatment (table 16). A thermo-mechanical treatment, applied to Hadfield steel, enables a grain size of the order of 10 µm, with a few percent of carbides to be obtained. Solution treatment conditions can be optimized to ensure the dissolution of carbides whilst limiting the grain growth. Tests have shown that carbides enter into solution at 1000°C and that the grain size 25 µm remained well below that obtained under normal industrial conditions (115).

Fine-grained alloys, stable when heated, are generally superplastic, and Hadfield steel presents this particularity. After hot rolling at 800°C, followed by cold rolling and annealing at 800°C, steel with 1.38% carbon, 12.6% manganese, and 0.07% silicon, had a grain size of 1 to 7 mm and a superplastic behaviour at 800°C at A deformed test piece at 800°C, at a rate of 1.67 10^{-4} s^{-1} in adverse test conditions, had 225% elongation at fracture (115).

Hadfield steels with high carbon contents and with a γ + carbides microstructure, under certain conditions of usage, have a higher abrasion resistance than the standard Hadfield steel. The combination of a fine grain and a uniform dispersion of globular carbides, obtained by thermo-mechanical treatment despite the localization of carbides in the grain boundaries, enables any toughness loss to be minimized.

	Manual Welding	Semi-Automatic Welding
Heterogenous Welding	18Cr-8Ni-6Mn (AWS A 5.4 E307) 20Cr-10Ni-3Mo (AWS A 5.4 E308 Mo)	Wire AWS Class A ER307, or A 5.9 ER308 Mo Flexed Cored Wire AWS CLass A 5.22 E307 T, or A 5.22 E308 MoT
Homogenous Welding (work-hardenable deposit)	13Mn-3Ni-Mo (AWS Class A 5.13 E FeMn)	Flexed Cored Wire 13Mn-3Ni-Mo with or without shield gas

Table 15 : Finishing Metals Used for Welding Hadfield Steels (111)

Chemical Composition wt%						Treatment	Hv	$E_{0.2}$ (MPa)	A %
C	Mn	Si	Cr	V	Ti				
1.10	13.6	0.4				TE 1100°C	200	436	52
						TTM	260	600	34
1.10	12.6	1.38	1.62	0.23		TE 1000°C	230	487	45
						TTM	435	800	20
1.19	12.0	0.64			0.105	TTM	205	760	28

Table 16 : Mechanical Properties of Manganes Austenites in the Quenched State, and after Thermomechanical Treatment. TTM = 70% reduction at 800°C

III. Heat Treatment

a) Solution treatments

Solution treatments should ensure the dissolution of unbalanced primary carbides which were formed during solidification, and those which were precipitated during cooling, and reduce as far as possible the chemical heterogeneity of the austenite resulting from dendritic segregation.

Pearlite begins to coalesce towards 700°C (after holding for one hour) and the pearlite's lamellar structure disappears towards 900°C ; a large amount of dissolution occurs at 1000 °C, but it is only at 1050-1100°C that complete dissolution can be ensured in the absence of carbide-forming elements. At this temperature the dendritic structure, revealed by macroetching, loses its contrast, indicating a certain homogenization of the austenite. However, disturbing phenomena can occur : the grain can begin to grow, and although the temperature is below that of the Fe-Mn-C eutectic, liquation can cause intergranular aggregates (C and P segregation give a melting point at 960°C.) The structure of the latter, at low melting point, has been identified as being composed of primary carbides and eutectic phosphides, and their formation is promoted by high phosphorus, carbon and molybdenum contents, and by the extent of the dendritic segregation (17) (69). Localized melting, observed particularly in the triple grain boundaries, extends into the primary boundaries and ends by forming a network when the solution treatment temperature increases.

Solution treatment is rather difficult, and precautions have to be taken. Taking into account the low thermal conductivity of these steels, heating should ensure a temperature as uniform as possible in the parts during increase in temperature, to avoid deformation and cracks : to do this, a homogenization treatment is used at times, at an intermediate temperature (3-5 hrs. at 600°C) (19). The solution treatment temperatures generally used for Hadfield and metastable austenitic steels are between 1050 and 1100°C, and 1150°C for chromium and vanadium grades (69) (19) (116).
The combination of extreme values of the heterogeneity of the furnace charge temperature, and the thermal oscillations due to the control system, should be situated between the temperature which enables carbide dissolution to be ensured, and the one at which liquation phenomena appear.

During atmospheric heating these steels undergo major decarburization. The depth of the decarburized layer can extend 2 to 4 mm (figure 90) and the carbon content of the surface layers can fall from 1.2% to 0.25% when decarburization in the mould is added to the decarburization in the furnace (117) (118).

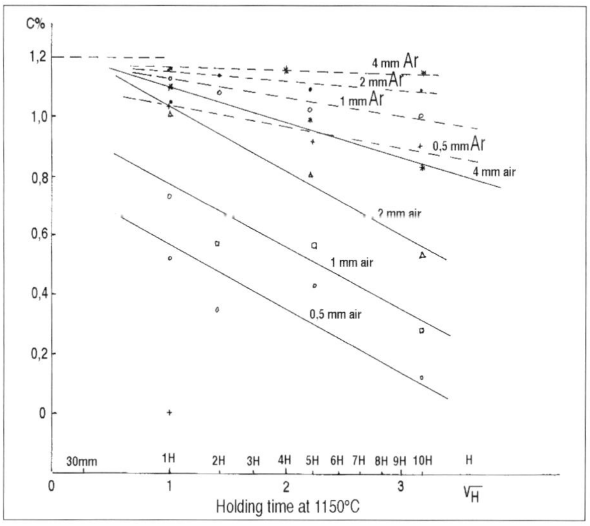

Figure 90 : Decarburization of Hadfield Steel at 1150°C (117) (5)

b) Treatments to improve machinability and grain size refinement

The pearlitic transformation of manganese austenites in isothermal conditions improves their machinability, and after re-austenization the grain size is refined. The pearlitic transformation kinetics of Hadfield steels are very slow, and one can count on only a partial transformation : even so it is difficult to operate in optimal conditions for decomposing the maximum amount of austenite.
The range of temperature of the pearlitic transformation is between 525°C and 625°C for the same grade of steel. The C curve is relatively closed, and the transformation kinetics slow considerably as soon as the treatment temperature varies from the preferred range. In industry one obtains at best 60% transformation, and more often between 20 and 40% after 16 hrs. of tempering at 550-600°C.

Machinability is improved after pearlitic transformation and after re-austenization a mixture of grain sizes is obtained, composed of coarse grain corresponding to the untransformed part of the structure, and fine grains which surround the latter.

Another process for refining grain size consists of work-hardening followed by a recrystallization treatment ; e.g., the grain size of Hadfield steel in the as-cast state, changes from a diameter of 0.125-0.250 mm to 0.063 mm after a 20% deformation followed by a recrystallization treatment of 45 min at 1050°C.

Metallographic examination has revealed that the initial dendritic segregation structure is unchanged by work-hardening and recrystallization (119), and in all probability this observation can also be applied to the grain refining obtained after solution treatment of a pearlitic structure. The grain refinement obtained by

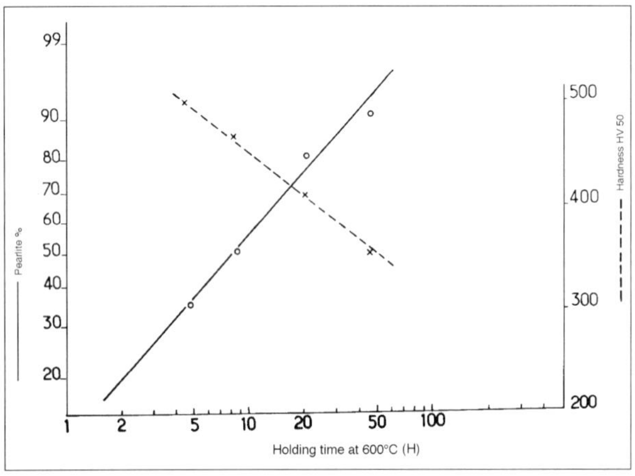

Figure 91 : Relationship between % Pearlite. Hardness, and Maintening Time at 600°C of a Metastable Austenitic Steel with 1.21%Cn 7.12% Mn, 0.93% C (Solution Treatment 30 min at 1050°C)

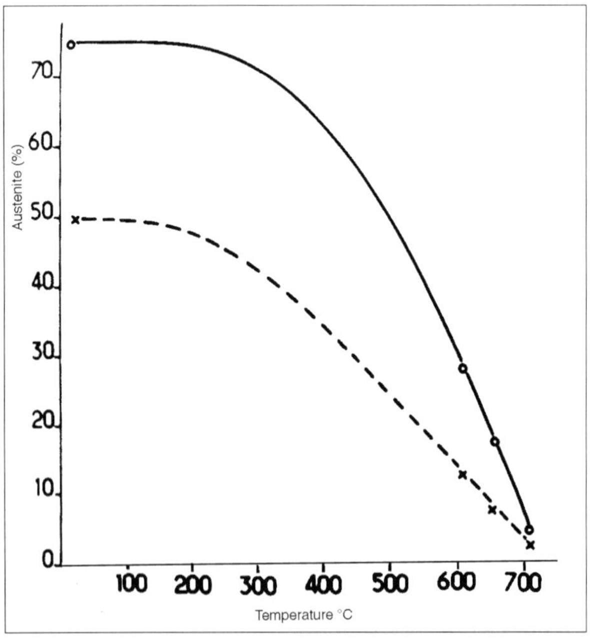

Figure 92 : Effect of Temperature (1 hr maintening) on % of Untransformed Austenitic of a Metastable Austenitic Steel Partially Pearlitized beforehand
Steel with 1.21% C - 7.12% Mn - 0.93% Mo
Preliminary Treatment
——— 30 min at 1050°C WQ + 600°C - 3hr30 (25% pearlite)
- - - - - 30 min at 1050°C WQ + 600°C - 9hr (50% pearlite)

following these procedures will not have any real effect on toughness except in so far as the embrittling effect of segregation is not dominant.

The pearlitic transformation kinetics of metastable austenitic steels are far more rapid than those of Hadfield steels, and all the austenite can be transformed into pearlite. For example, the microstructure of the steel with 1.2% C - 7.2% Mn - 0.93% Mo contains 90% pearlite after holding for 48 hrs. at 600°C (figure 91). Treatment time can be shortened, after partial transformation of the austenite (25 to 50% pearlite, tempering for 3.5 hrs to 9 hrs. at 600°C) by proceeding with a second treatment for 1 hr. at 700°C (figure 92). The austenite not transformed during pearlitic transformation becomes destabilized simultaneously during the treatment, and then transforms rapidly after heating at 700°C into a very fine ferrite, globular carbide and untransformed austenite aggregate.

The machinability of metastable austenitic steels treated in this way is the equivalent of that of hypereutectoid tool steels. The grain size of the steel given as an example changes from an ASTM grain index of 2-3 in the as-cast state, to 6-7 after treatment.

c) Treatments for carbide precipitation hardening

Manganese austenites are generally not hardened by carbide precipitation in the matrix, precipitation occuring most often in the grain boundary, thereby embrittling the material. However, treatment cycles enable the formation of a dispersion of carbides in the austenite. The hardening obtained in this way cannot be compared to the hardening resulting from a coherent precipitation but it has a marked beneficial effect on yield strength and abrasion resistance, but on the other hand leads to a decrease in toughness.

The cycle shown in figure 93a consists of a solution treatment, followed by a pearlitic transformation in isothermal conditions, followed by partial dissolution of the pearlite's cementite lamellae. The cementite lamellae coalesce during dissolution, and small globular carbides are obtained which are quite well distributed in place of the former pearlite colonies, the latter having reverted to austenite form.

Figure 94 (see p. 60) shows the effect of spheroidization-dissolution treatment temperature of the carbides on the mechanical properties of austenitic steels with 1.2% carbon, varying manganese contents, and with or without the addition of molybdenum.

After a complete transformation (to α + carbides) of metastable austenitic steels, followed by a solution

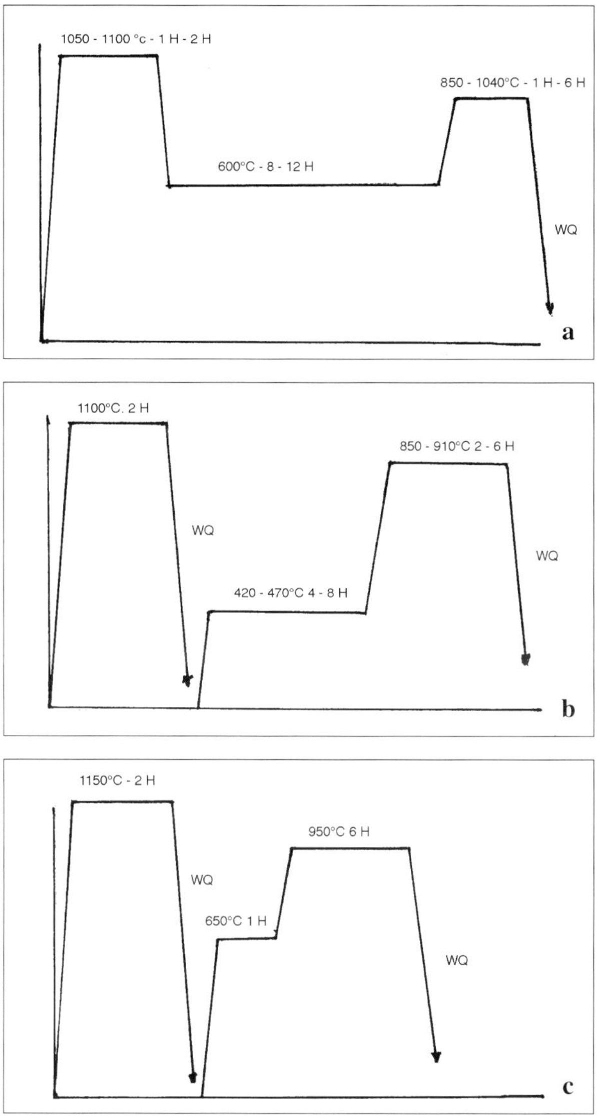

Figure 93 : Thermal Treatment Cycles to Obtain Globular Carbide Dispersion in the Matrix
a) Pearlitic Transformation and Dissolution-Spheroidization (12 Mn - 2 Mo Steel) (75)
b) Precipitation of Carbides as Platelets and Dissolution-Spheroidization (classic Hadfield Steels) (122)
c) Precipitation of Vanadium Carbides (0.8% C - 13% Mn - 1% V) (128)

treatment, a mixed structure (γ + carbides) or ($\alpha + \gamma$ + perlite) is obtained, according to the quantity of carbon in solution. Table 17 gives, as an example, the Ms temperature of a steel with 1.2% C - 7.1% Mn - 0.9% Mo, according to solution treatment conditions. It can be estimated that 0.4% carbon dissolved in the austenite gives an Ms point at 188°C and a martensitic structure containing 20% austenite and 12% carbides. 0.7% carbon in the austenite lowers the Ms temperature to room temperature, and the austenitic structure after cooling contains 7.5% carbides.

The ($\alpha + \gamma$ + carbides) structures can benefit from hardening after tempering at around 500°C when the amount of retained austenite is sufficient to compensate for the softening of the previously formed martensite during tempering.

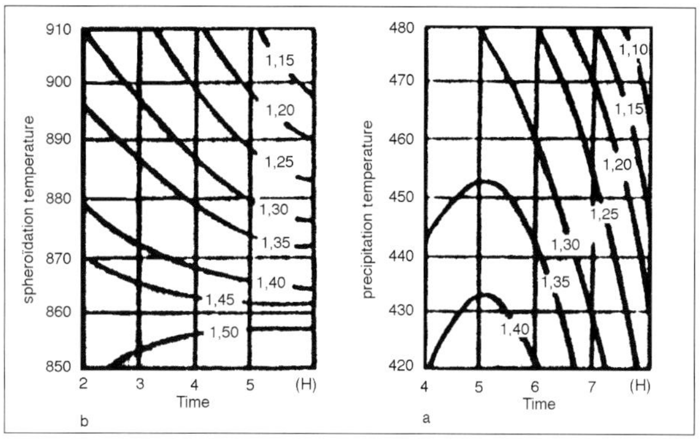

Figure 95 : Influence of Carbide Precipitation and Spheroidization Treatments on Abrasion Resistance of Hadfield Steel with 1.10% C, 13% Mn (122)
(Quenching at 1100°C ; Precipitation of carbides as platelets 420-470°C - 4-7 hrs, spheroidization 850-900°C - 2-5 hrs (abrasion tests on abrasive bands) Wear resistance is measured by the radio : Wear in treated state/Wear in quenched state at 1100°C)

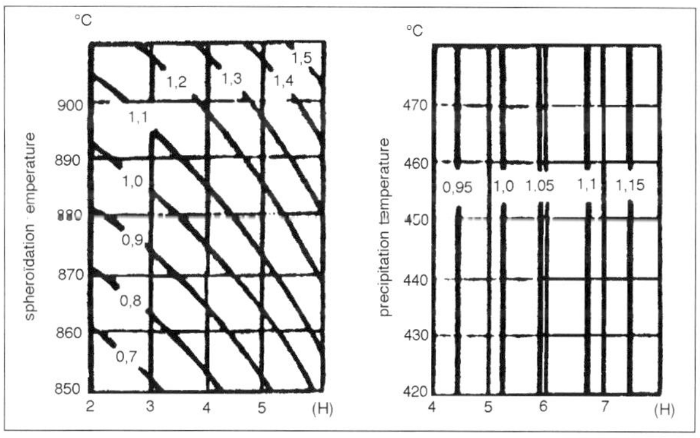

Figure 96 : Influence of Carbide Precipitation and Spheroidization Treatments on the impact of Hadfield Steel with 1.10% C, 13% Mn (122)
Quenching at 1100°C
a) Carbide Platelet Precipitation 420-470°C, 4-7 hrs
b) Spheroidization 850-900°C, 2-5 hrs
(Impact strength in MJm^{-2} - Test specimen type unspecified, probably Mesnager)

Solution Treatment			
T (°C)	**t (mn)**	**M_S (°C)**	**γ %**
760	30	68	59
760	120	47	74
800	120	28	92
850	30	< 20	100

Table 17 : Effect of Solution Treatment Conditions on Ms Temperature, and Austenite %, of a Pearlitized Metastable Steel (1.2% C, 7.2% Mn, 0.93% Mo)

Table 18 shows the chemical composition of other steels as 6-1 intended for industrial applications, in the austenitic or martensitic state (120) (121).

The second method for obtaining carbide distribution differs from the first merely by a lower isothermal transformation temperature (450-480°) as against 600°C, which gives little if any pearlite, forming instead carbide platelets (figure 93b). After spheroidization-dissolution treatment of the carbide platelets, the spheroided carbides are uniformly

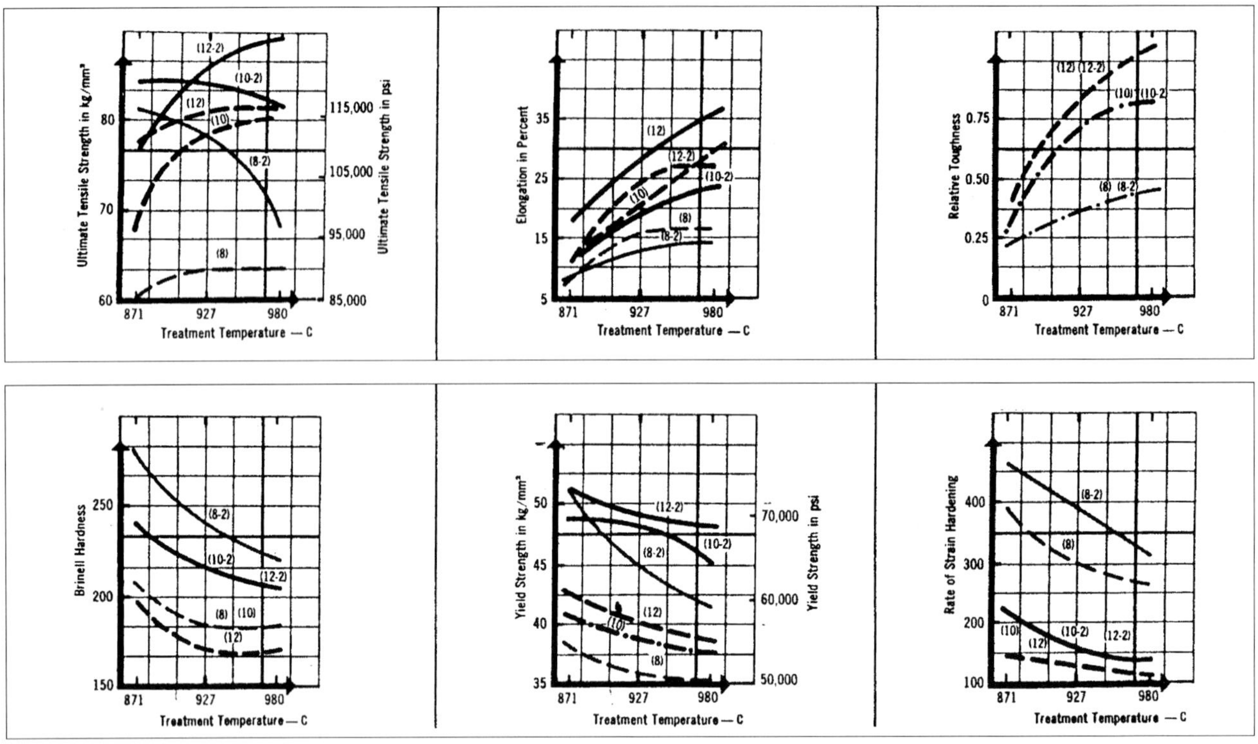

Figure 94 : Influence of Spheroidizing Temperature on Tensile Properties of Austenitic Steels having Various Manganese Contents, with an without Mo Additions (75)

distributed in the austenitic matrix. This method is suitable for standard Hadfield steels, but does not apply to those which contain elements such as molybdenum, which inhibit the formation of carbide platelets. The effect of carbide precipitation treatment conditions and spheroidization-dissolution on the impact and abrasion resistance of a Hadfield steel with 1.1% C is shown in figures 95-96.

Name	Chemical Composition %						
	C	Mn	Si	Cr	Mo	V	Co
Calloy (120)	1.1	5.5	1.0	1.7	1.5		1.8
7 G6F2 (121)	0.7-1.3	5.4-10	0.2-1.0			0.5-2.7	

Table 18 : Controled Structure Steels

The addition of vanadium to manganese austenitic steels leads to carbide precipitation in the matrix after aging. The non-magnetic steel with 0.45% C- 18% Mn - 3.5% Cr - 1.25% V is hardened by precipitation of fine vanadium carbides (150Å) in the matrix after aging of 4 to 8 hrs. at 650°C (123). An increase in the carbon content of manganese austenites in the presence of vanadium, promotes intergranular carbide precipitation which embrittles the material, as shown by the analysis of the results obtained by different authors (124-127). Limiting the carbon content to 0.8-1.0%, and a two-stage aging treatment after a solution treatment at 1150°C (figure 93c), minimizes the formation of intergranular carbides, and gives a uniform precipitation of small vanadium carbides (12) (128). The aim of the first stage of the aging treatment (1 hr. at 650°C) is to precipitate vanadium carbides

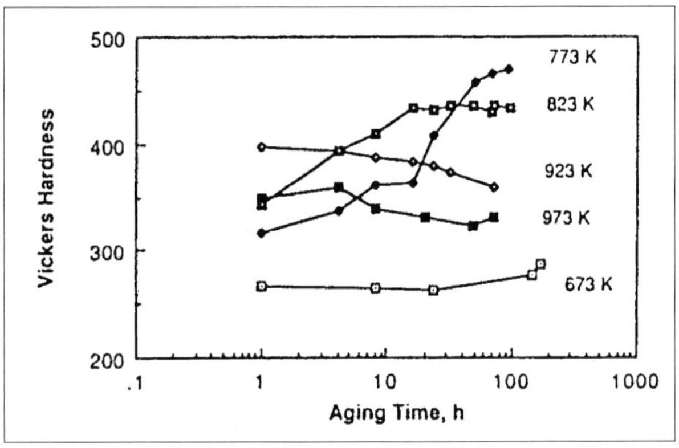

Figure 97 : The Variation of Hardness as a function of Aging Temperature and Time (1% C - 30% Mn - 1% Si - 10% Al) (23)

onto the stacking faults formed during heating and tempering. The second stage (6 hrs. at 950°C) ensures the growth of carbides, the latter having reached critical nucleation size during holding at 650°C. The matrix, originally containing relatively little carbide, is depleted in carbon by vanadium carbide precipitation, and the austenite becomes metastable.

The conditions of hardening of the grade under consideration rely solely on test pieces of thin section, taking into account the vanadium carbide precipitation kinetics during quenching. The limit of the section size susceptible to the treatment has yet to be specified.
A steel with 1.0% C - 30% Mn - 1% Si - 10% Al hardens by spinodal decomposition and carbide κ precipitation at a temperature of between 450 and 700°C (figure 97).

IV. Machinability of Austenitic Manganese Steels

In general machining operations carried out on austenitic manganese steels aim for the majority of them to realise plane surfaces or revolution shapes through turning (traversing, broading, rectifying) milling, surfacing, boring and grinding. This chapter written by CETIM (192) reviews the machinability of manganese austenitic steels together with precautions to be taken with their machining with tools.

a) Machinability

Machining by chips removal is a shear deformation of the material which induces an important superficial work hardening, for instance the superficial hardness measured on an Hadfield steel after milling was 540-590 HV1 and the depth of the work hardened zone was 0.25 mm. Studies carried out on chip formation on this type of material have shown that cutting stress was important. Stress decreases when the cutting speed increases and then stabilises (in the case of turning this is obtained at 20m/mn).

Cutting temperature is high and measurements show that the temperature is the same on the rake face and on the cutting edge which is the opposite of what occurs to classical steel grades where the hottest zone is the chip's friction zone on the rake face. This phenomena can lead with high speed tool steel to a creep deformation of the cutting edge. Variable strains in the primary shear zone of the tool give instabilities of the shape of this chip (making irregular saw teeth) and consequently a chatter is induced by the chip formation process.

The wear of the cutting tool through the addition of these effects can be characterised as follows:
The main wear factor is the sticking of the tool on the clearance face due to the low cutting speeds dictated by the behaviour of this material.
The hardening caused by the first going over, together with diffusion and oxidation wear, results in a deep flank wear of the tool at the depth of cut which is named notch wear.

There is generally no cratering wear on the rake face of the tool as the plastic deformation of the cutting edge resulting from the high temperature occurs first.

Machinability of manganese austenitic steels is low compared to other steels. The machinability index measured by the ratio of the cutting speeds used, (for turning) is about the following:

basis is medium carbon steel
Austenitic stainless steel = 50%
Manganese austenitic steel=10%

Actions to improve the machinability of this material outside of the constant improvements of the cutting tools themselves (grade and geometry) can be considered for steel elaboration and for cutting processes.

It can be noted in the technical literature interesting results obtained with grades designed for a better machinability. In non-magnetic grades, which do not have a good wear resistance, sulphur and calcium additions allow the increase of machinability and decrease the cutting temperature. Sulphur combines with manganese to form a sulphide which acts as a lubricant and at the same time reduces the toughness of the material favouring the breaking up of chips.

A calcium addition in combination with residual aluminium and silicon, forms non abrasive oxides which reduce the tool wear. For example when turning without a lubricant with a M 20 type tool, and a cutting speed of 40 m/min the cutting temperature with a 15% manganese steel, drops 100° C with an addition of 0.04% sulphur and 130° C when 0.05% sulphur and calcium are added. This lowering of the cutting temperature is shown in another example concerning drilling with a twist-drill, the tool life is considerably improved for constant cutting conditions. An increase of the sulphur addition from 0.02% to 0.04% can double the life. In the case of face milling the same sulphur increase improves the tool life by 50% and another 50% when the steel is also calcium treated.

On the other hand it is also possible to improve the machinability by an isothermal treatment at 650° which forms perlite. This method requires after machining a new solution treatment followed by water quenching with unavoidable deformation of the part.

This technique is rather easy to apply with small parts when machining has to be carried out in difficult accessible zones (blind hole, groove, drilling, end milling). With large parts it is more difficult and care should be taken in order to minimise deformation.

b) Precautions for Machining

To lessen difficulties when machining this material, care must be observed in order to reduce deformations, impacts, stresses and cutting temperature which are causes of rapid wearing of the tool.

- Avoid tool stops and machining resumptions with tool friction on the part, for example:
  - When turning, avoid prolonged friction of the tool against the shoulder,
  - When stopping traverse feed.
  - When milling avoid friction of the surfacing tool on the part when reference taking.
  - When drilling avoid excess temporising before lifting the drill to remove chips.

- Avoid operating on zones work hardened by previous machining operations.
  - Feed and depth of going over must be sufficient so that the cutting edge acts under the work hardened layer.
  - Rimming after drilling,tapping should be avoided.

- Rough holes should be considered if possible, when casting, as drilling is a difficult operation. The highly negative geometry of the drill-end does not facilitate for tool penetration.

- Choose tools with sharp edges, as a rounded edge increases the workhardening and machining difficulties, choose a positive resultant to reduce stress.

- Use rigid machines tool, tool holders, and part holders in order to avoid vibrations caused by high stresses and chattering, induced by chip formation on this material.

- Use lubricant with extreme pressure additives.

c) Tools Choice and Cutting Conditions for a Steel 1% C - 12-14% Mn

The more common machining operations are turning, milling,drilling and grinding.

c. a) Grade and Geometry

Generally speaking cutting geometries with a negative resultant are more resistant but generate more stress which means more work hardening and also generates a higher temperature at the tool end. Geometries with a positive resultant are preferred, albeit more vulnerable to temperature they reduce work hardening.

Concerning high speed steels a higher resistance will be obtained by using a cobalt high speed steel (minimum 8%) which holds a better temperature, a higher wear resistance is ensured for tool grades for high speed steels. With a cobalt grade (minimum 8%) which has better hot properties.

The use of carbide tools is preferable but the choice of grade is critical. The hardest grades wear and plastic deformation resistant are not satisfactory. Due to low cutting speeds, stress and chattering induced by cutting lead to chipping and flaking (especially with positive geometries) that is to random tool life.

It is better then to have a tougher grade and this is why, that somewhat softer grades which are a better compromise, as is the case with ISO M 20 or K20 are chosen to secure more regular tool lives. It can also be noted that the use of covered carbide does not always maintain a good sharp edge.

c.b) Turning

Cutting geometries associated with carbide containing grades M20-K20 are of the negative/positive type, this means a negative positioning of the insert on the tool holder to allow for the use of a strong negative insert (required because of the stress involved), but as a positive geometry chip breaker, not to control a correct chip breakage, when cutting speed is over 15m/min, but to reduce the workhardening effect.

Data below is a synthesis of machining experiences and information supplied by tool manufacturers. It is (the beginning data) for tool life around 30 minutes, as tool behaviour is highly dependent on the relative performances of each tool grade and each material to be machined.

Turning Single Point

ap	f	vc	Tool Grade
1	0.2	45	M20-K20
3	0.4	30	M20-K20

Note
ap = depth of cut (mm)
f = feed per revolution (mm)
vc = cutting speed (m/min)

c.c) Milling

For face milling geometries used can be milling-cutters with strong rounded inserts associated to M20 or K20 type tool grade but with a chip breaker giving a positive resultant cutting geometry.

Tool lives are variable according to performances of grades supplied by carbide makers and data shown below is only starting range to be applied.

Face Milling

ap	fz	vc	Tool Grade
0.5	0.15	30-40	M20-K20
2	0.25	20-30	M20-K20

Note
ap = depth of cut (mm)
fz = feed per tooth (mm)
vc = cutting speed (m/min)

c.d) Drilling - Boring

When drilling with high speed steel twist drills it is advisable to use drills with 8% cobalt which have a good hot strength (Type 2918).
The following is the recommended high speed steel drill geometry:

- Short drill with a reinforced web for the highest rigidity.
- Thinning down of the web end to ease drill penetration.
- Low rake angle
- End angle of 130°

When drilling with a carbide drill, tool grades are grade ISO K 10 type, in some cases covered, allowing about an increase of 30% of the cutting speed.

Data below was completed with experimental data.

Drilling
d = tool diameter

vc	d = 6mm	d = 12mm	Tool Grade
5	f = 0.06	f = 0.12	high speed 2918
25.40	f = 0.03	f = 0.06	Carbide K 10

Boring with carbide tool, vc = 30-70 m/min
f = 0.15 for d = 20 mn and f = 0.25 pour d = 65 mn
Note f = feed per revolution (mm)
vc = cutting speed (m/min)

c.e) Grinding

Casting shops specialising in austenitic manganese steel production use for trimming or grinding surfaces or the inner of cylinder, alumina or silicon carbide wheels with average grain size (46 or 54), medium grade (J or K) a rather open structure (12), and a fused binder.

c.f) Hot Machining

Heating techniques used for the zone where chip forms, are electric currents, oxyacetylene flame, electric arc, but also a plasma torch considered the best for energy yield. To make a choice among these heating processes compared to traditional machining it is necessary to undertake a technico-economical analysis taking into account, machining cost, investment and utilisation cost, the possibility to move the equipment on various machines, environmental constraints as well as technical results. Industrial results for turning or milling show chip removal speeds increased by a factor 10 with better tool life and improved surfaces when carbide and ceramic tools are used. The following example is an illustration of possible gains :

Conventional milling : cutting speed Vc = 20 m/min
tool life T = 10 min
Hot milling : Vc = 200 m/min

Part 3 - Properties

I. ***Physical Properties***

II. ***Corrosion and Oxidation resistance***

III. ***Mechanical Properties***

IV. ***Contact Deformation***

V. ***Armour Plates***

I. Physical Properties

a)Crystal lattice parameter, density

The austenite's crystal lattice parameter (a) is given by the following equation :

a (Å) = 3.58404 + 0.0357 (At % C) + 0.00418 (At % N) + 0.00014 (At % Mn) + 0.00143 (At % Cu) + 0.00012 (At % Ni) + 0.00009 (At % Cr) + 0.00343 (At % Mo) + 0.0002 (At % V) - 0.00293 (At % Ti)

The correlation index of this equation is 0.94 (129).

Density at room temperature : (88) (110) (131)	($g\ cm^{-3}$)
Hadfield steel (1.2% C - 13% Mn)	7.87
0.6 C - 14 Mn - 2 Ni - 2 Cr - 0.02 N	7.94
0.6 C - 30 Mn	7.73
AISI 304	8.03

b) Electric and Magnetic Properties

1. Resistivity ($\mu\Omega.cm^{-3}$) (88) (110) (108) (131)

Hadfield steel and manganese austenite with lower carbon content

An empirical relation between the electric resistivity (ρ) and the chemical composition of Mn Cr Ni austenitic steels gives the resistivity of manganese austenites with low carbon contents :

$$\rho = 36.0 + 1.27\ Mn\ \% + 0.66\ Cr\ \% + 1.62\ Ni\ \% + 5.90\ Si\ \%\ (\mu\Omega.cm^{-3})\quad (130)$$

Temperature Range (°C)	-183	0	20	100	200	400	600	800
Hadfield Steels	53	66		76	84	99	110	121
0.6C - 14Mn - 2Ni 2Cr - 0.002N			68.4					
0.7C - 15Mn - 1Ni			71.0					
0.6C - 30Mn			84.8					

The composition domain of the equation's definition is given by :

C % : 0.09 - 0.16, Si % : 0.1 - 5.2,
Mn % : 5.0 - 26.0, Cr % ≤ 15.0, Ni % ≤ 6.0
N % : 0.049 - 0.086, Nb % ≤ 0.7

2. Magnetic Permeability

It is generally accepted that the magnetic permeability of manganese austenites, used in non-magnetic applications, should be less than 1.02. Manganese austenites which are in the γ and γ + ε domains in the quenched state at 1100°C amply fulfill this criterion (figure 98). The extent of the composition domain which meets this criterion is practically

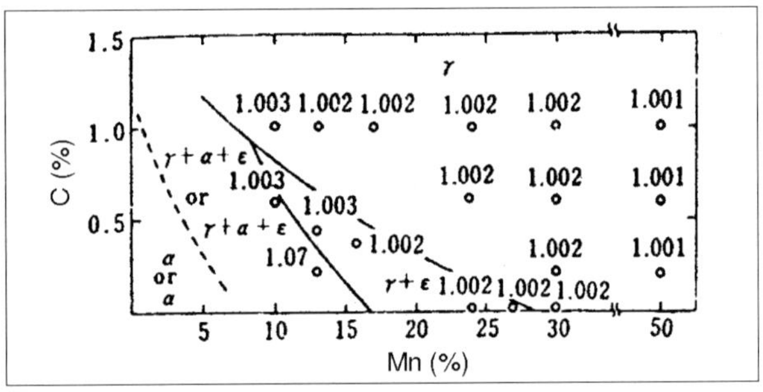

Figure 98 : Magnetic Permeability of Manganese Austenites at the Quenched State at 1100°C (130)

unaffected by cold deformation (figure 99). The precipitation of M_3C, $M_{23}C_6$, M_6C and MC carbides into the austenite, does not change magnetic permeability, the latter increasing with the appearance of pearlitic transformation (figure 100).

c) Thermal properties

1. Expansion

The expansion coefficient α, between 0 and 100°C, is given as a function of chemical composition and of the percentage of martensite ε, by the equation :

$$\alpha = 19.72 + 2.32\ C\ \% + 1.39\ N\ \% - 0.31\ Mn\ \% + 0.028\ (Cr\ \%)2 - 0.41\ Cr\ \% + 0.14\ \varepsilon\ \%$$

($10^{-6}.°C^{-1}$)

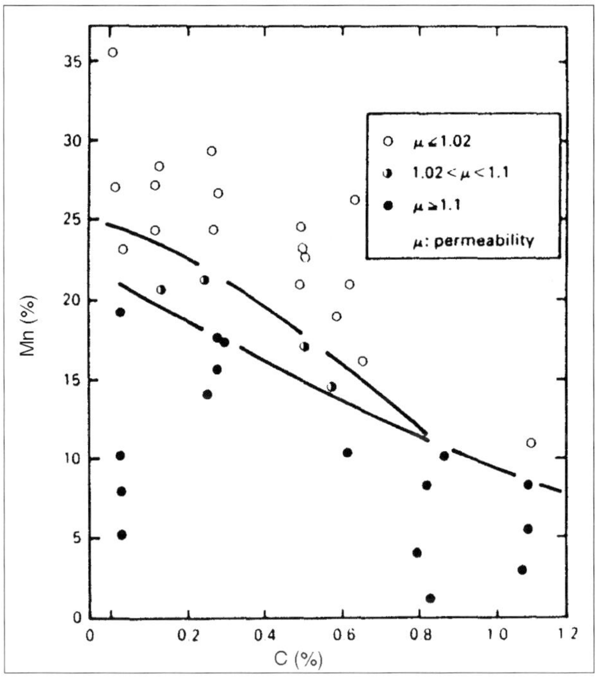

Figure 99 : Magnetic Permeability of Manganese Austenite after 50% Cold Reduction by Rolling (131)

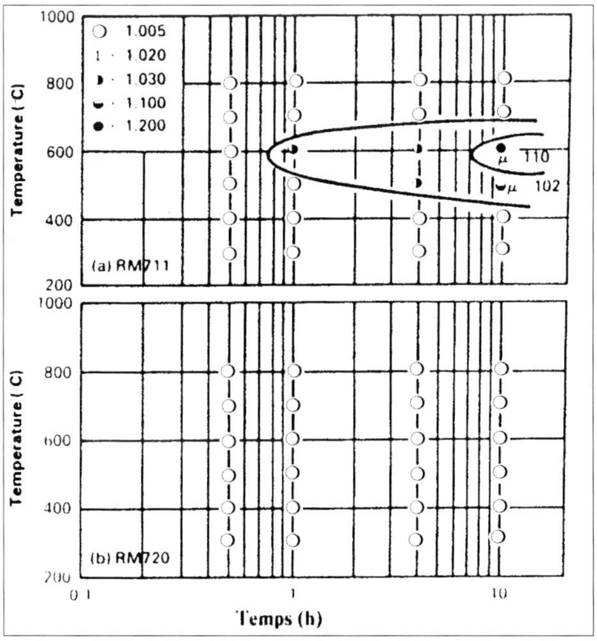

Figure 100 : Effect of Isothermal Holding on Magnetic Permeability of Manganese Austenitic Steels Quenched at 1100°C (132)

Steel	C%	Mn%	Ni%	V%
RM 711	0.7	15.0	1.0	
RM 720	0.7	15.0	1.0	0.6

The correlation coefficient of this equation is 0.933, with composition range of :

C % : 0.01 - 0.9, N % : 0.015 - 0.26,
Si % : 0.04 - 0.46, Ni % : 0 - 1.2, Cr % : 0.01 - 7.01,
Mo % : 0 - 0.04 (108)

A simplified equation taking into account only the manganese and carbon contents has been suggested :

$\alpha = 17.66 + 3.82\ C\ \% - 0.22\ Mn\ \%\ (10^{-6}.°C^{-1})$

in a definition range between 10 and 50 % manganese and between 0.02 and 1 % carbon (131).

These equations show that it is possible to adjust the expansion coefficient according to manganese and carbon contents. An expansion coefficient can be obtained which is less than that of low carbon constructional steels, with high manganese and low carbon content.

The small expansion coefficient of high manganese, high carbon austenitic steels is attributed to the high value of the volume magnetostrictive effect, below the Néel temperature which cancels the volume variation due to the atomic lattice vibration.

The expansion coefficients of Hadfield steel and of lower carbon manganese austenite ($10^{-6}.°C^{-1}$) (88) (110) (131) are shown below.

Temperature Range (°C)	0/100	0/200	0/400	0/600	0/800	0/1000
Hadfield Steels	18.0	19.4	21.7	19.9	21.9	23.1
0.6C - 30Mn	11.0					
0.6C - 14Mn - 2Ni 2Cr - 0.002N	16.2					
0.6C - 18Mn - 1Cr Ni 1 V 0.03 N	13.0	20.0		22.0		
AISI 304	17.3					

2. *Heat Capacities*

($cal.g^{-1}.°C^{-1}$) (88) (110)

Temperature Range (°C)	25	50/100	150/200	350/440	550/600	750/800	950/1000
Hadfield Steels		0.124	0.135	0.145	0.168	0.155	0.161
0.6C - 14Mn 2Ni - 2Cr - 0,002N	0.118		0.127	0.135	0.141	0.153	0.158
AISI 304			0.125	0.132	0.138	0.147	

3. *Thermal Conductivity*

($cal.cm^{-1}.s^{-1}.°C^{-1}$) (88) (110)

Temperature Range (°C)	0	25	200	400	600	800	1000
Hadfield Steels	0.031		0.039	0.046	0.052	0.056	0.061
0.6C - $14Mn^2$ 2Ni - 2Cr - 0,002N		0.0360	0.0399	0.0442	0.0542	0.0606	0.0694
0.6C - 30Mn		0032					
AISI 304			0.0405	0.0469	0.0534	0.0599	

d) Mechanical Properties

1. Young's Modulus at 20°C in MPa

Hadfield Steels	203 000-205 000
0.56 C 18 Mn 5.5 Cr	193 000
0.6 C 14 Mn 2 Ni 2 Cr 0.02 N	203 170
0.6 C 30 Mn	173 000

Figure 101 shows the effect of temperature on Young's modulus of Hadfield steel (dynamic bending measurement). Young's modulus is lower when the phosphide eutectic gives a network in the grain boundaries.

2. Internal friction

Internal friction of Hadfield steels increases with an increase in temperature, and the increase is very rapid above 600°C. The curve shows numerous anomalies up to 800°C which can be attributed to phosphorus content and austenitic transformation.
Elastic wave attenuation during propagation, via energy dissipation due to internal friction, limits the detection of internal defects by ultrasonic inspection. The ultrasonic attenuation index increases with an increase in grain size and frequency (figure 102). The signal/background noise ratio corresponding to a critical diameter defect should generally be above 3, and the frequency should be as high as possible (figure 103).

Figure 101 : Temperature Dependence of the Modulus of Elasticity of Cast Steels 1.1 C, 13 Mn with 0.016 and 0.062% P (133)

Figure 102 : Effect of Grain Size and Frequency on Attenuation Coefficient of Ultrasound (0.6 C - 14 Mn - 2 Ni - 2Cr - 0.002 N) (110)

Figure 103 : Detectability of Artificial Flaw (110)

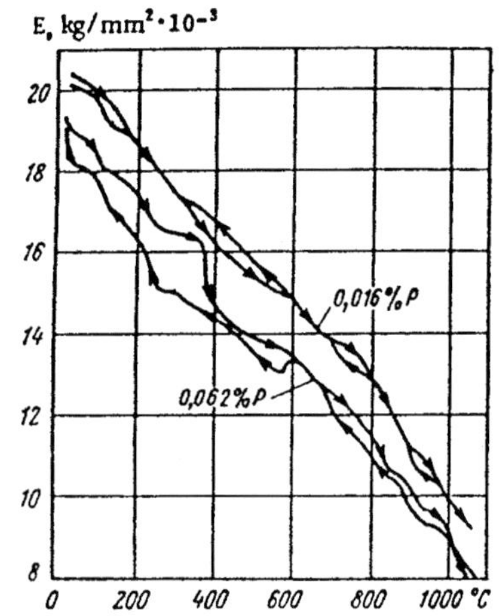

Figure 101

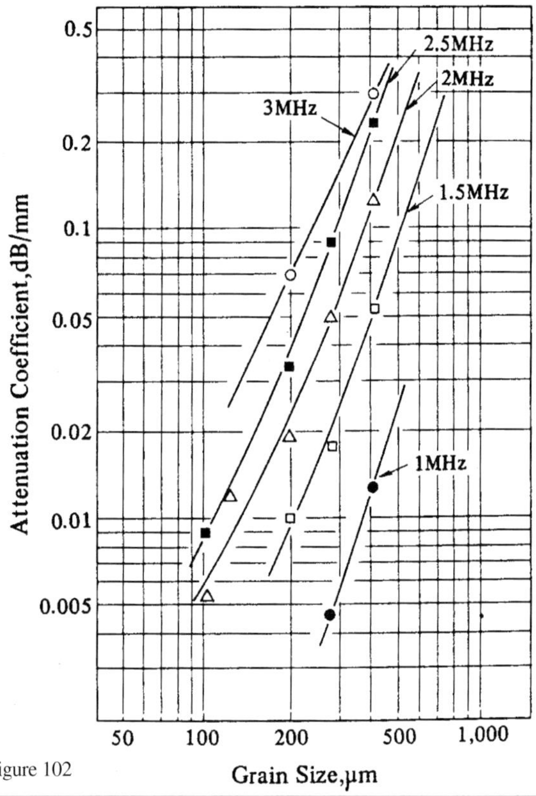

Figure 102

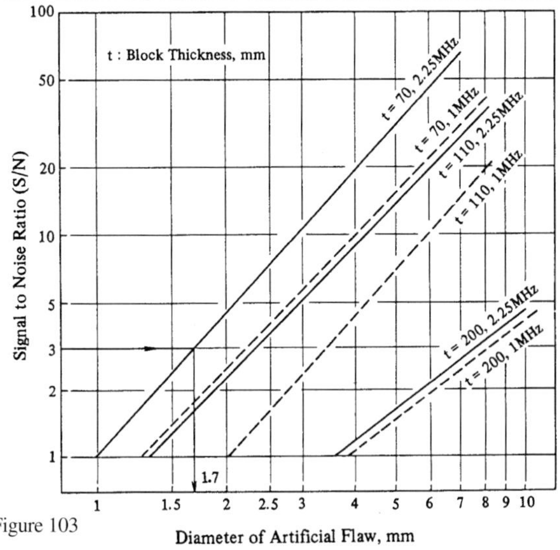

Figure 103

II. Corrosion and Oxidation Resistance

Manganese austenites have a low corrosion and oxidation resistance and to rectify this chromium is added, or aluminium with silicon. The effect of chromium on corrosion resistance in oxidizing and chlorinated environments is well known : steels containing C < 1%, Mn 5-10%, Cr 14-16% and Mo 0-3.0% were studied with a view to providing abrasion resistance in hostile environments (24) (139). Steels with 1% C 30% Mn 1% Si 10% Al have an excellent general corrosion resistance, as well as stress corrosion resistance in seawater, and excellent cavitation resistance (figure 104). This grade has been suggested as a replacement for brass for boat propellers (134) (135).

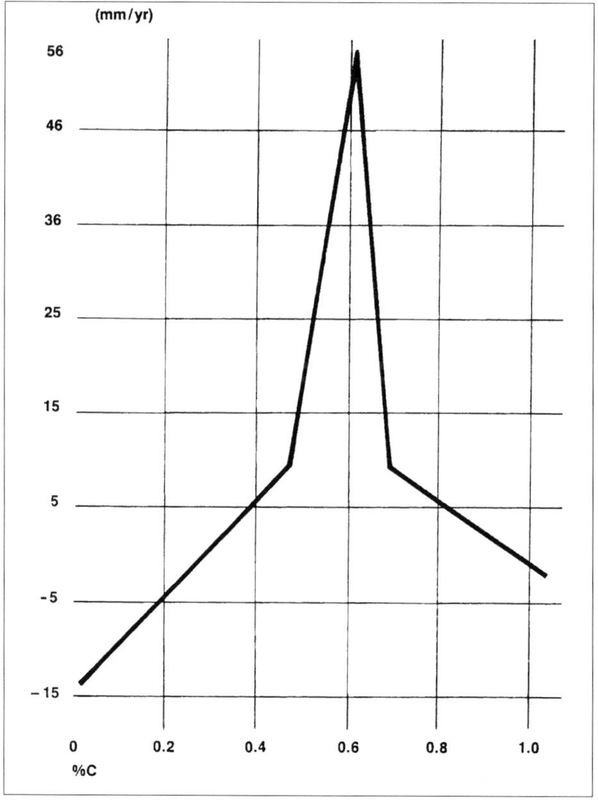

Figure 104 : Effect of Carbon Content on the Corrosion Rate of Steel (30 Mn - 10 Al - 1 Si) in sea Water (134)

The oxidation resistance of manganese austenites increases in a ratio of 1 to 10 when the chromium content changes from 5 to 10% (figure 105). During tempering between 500 and 700°C, chromium carbides precipitate into the grain boundaries of

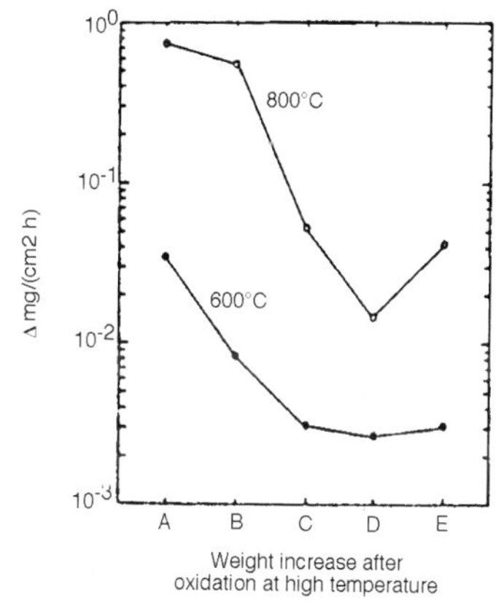

Chemical compositions of the specimens

	C	Si	Mn	P	S	Cr	Ni	V	N
A	0.56	0.46	17.7	0.017	0.002	5.32	0.96	0.51	0.031
B	0.60	0.51	18.3	0.018	0.009	5.91	1.01	0.97	0.025
C	0.58	0.51	18.3	0.020	0.009	10.58	1.02	1.00	0.026
D	0.58	0.55	18.0	0.019	0.009	10.84	1.00	1.90	0.027
E	0.56	0.64	18.1	0.020	0.008	10.56	1.02	0.98	0.091

Figure 105 : Oxidation Resistance at High Temperature of Austenitic Mn-Cr Steels (136)

austenitic manganese steels, which contain more than 3% chromium (figure 106). The material becomes brittle after cooling and susceptible to stress corrosion. A decrease in carbon content accompanied by an increase in chromium content considerably reduces intergranular carbide precipitation, and in consequence reduces embrittlement and susceptibility to stress corrosion. Non-magnetic steel containing 0.55% C 18% Mn 4% Cr used for retaining rings is embrittled and is susceptible to stress corrosion after tempering for 24 hrs. at 500°C. With a carbon content of 0.1% and a chromium content of 18%, there is very little carbide precipitation and, after a sensitizing treatment, toughness is little changed (108) and the material is not susceptible to stress corrosion in various media (138).

Manganese austenitic steels used for combustion engine valves and also for titanium forging dies, form a family of commercial steels (table 3). The austenitic structure of these steels, containing 21% chromium

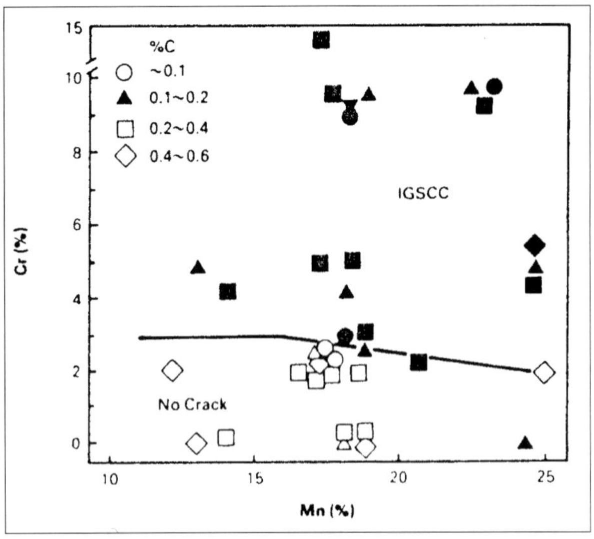

Figure 106 : Influence of Mn and Cr Contents on Stress corrosion Susceptibility of Steels with High Mn Content (108)
Thermal Treatment : 8 hr at 650°C and air cooling. Stress corrosion tests : Immersion in Synthetic sea water at 50°C for 720 hrs.
IGSCC : Intergranular stress corrosion

with relatively low contents of carbon (0.5-0.7%) and manganese (6-11%) is maintained by the addition of nitrogen and sometimes nickel. These steels have excellent hot corrosion resistance against the combustion gases of petroleum products. High manganese austenites have a higher hot corrosion resistance to sulphurous atmospheres than nickel austenites.

The oxidation resistance of steel with 0.93% C, 31% Mn, 7.5% Al, 1.3% Si, is slightly lower at 700°C than that of the stainless steel AISI 304. This resistance deteriorates at higher temperatures and at 800°C the oxidation resistance of the 304 steel is clearly higher (figure 107). The toughness of this steel is very low after holding above 550°C (23).

Figure 107 : Oxidation Characteristics of Cr-free Austenitic Steel (Fe-0.93 C - 31 Mn - 7.5 Al) with and without 1.3% Si at 700°C and 800°C (139)

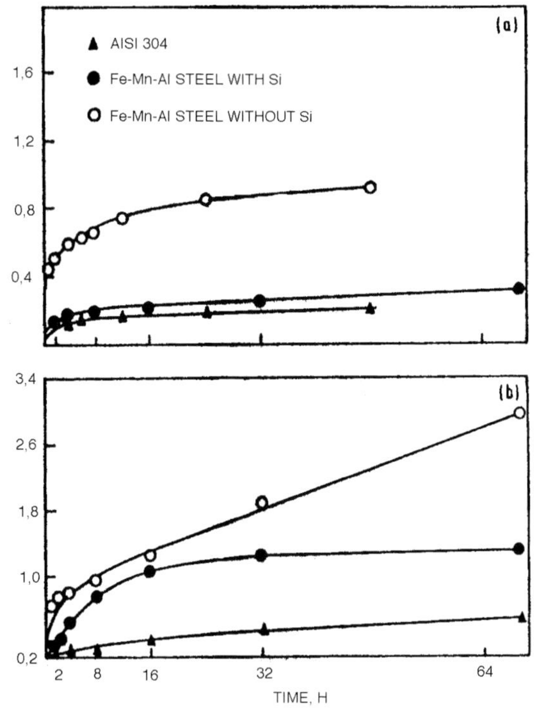

(a) 700°C ; (b) 800°C

III. Mechanical Properties

a) Yield strength

The yield strength of manganese austenitic steels obeys the Petch equation :

$$\sigma_{0.2} = \sigma_i + \kappa\, d^{-1/2}$$

which relates the yield strength $\sigma_{0.2}$ to grain size (d) (figure 108) (140) (71).

The relation :

$$\sigma_{0.2} = 299.2 + 135.7\ C\% + 505.2\ N\% + 111.5\ V\% + 24.3\ Mo\% + 9.1\ Cr\ \% - 4.6\ Mn\%\ (MPa)$$

established in the definition interval :

0.01 to 0.80 % C, 0.05 to 0.27 % N, 0.01 to 0.88 % V, 0.01 to 1.9 % Mo, 0.04 to 3.1 % Cr and 14.8 to 35.3 % Mn,

with a correlation coefficient of 0.92 (141), and the results given in table 19, show that the elements which produce the greatest increase in yield strength are nitrogen, carbon and vanadium, followed by titanium, molybdenum, chromium and aluminium. Nickel has no effect, and manganese has a slightly negative effect.

An increase in yield strength due to solution elements, shows the effect of their atomic diameters on the lattice parameters of the austenite. Figure 109 shows that increases of the yield strength and of the lattice parameters of manganese and nickel-chromium austenites, due to the addition of 1 at % alloying elements, can be positioned on a single curve ; however, two elements are the exceptions to this rule : for a lattice of the same size, nitrogen and vanadium produce an increase in yield strength of manganese austenitic steels which is higher than that observed for other alloying elements. Then again, vanadium and carbon do not have the same effect on the Petch equation. An increase in carbon content displaces the curve $\sigma_{0.2}$ - $1/\sqrt{d}$ towards higher yield

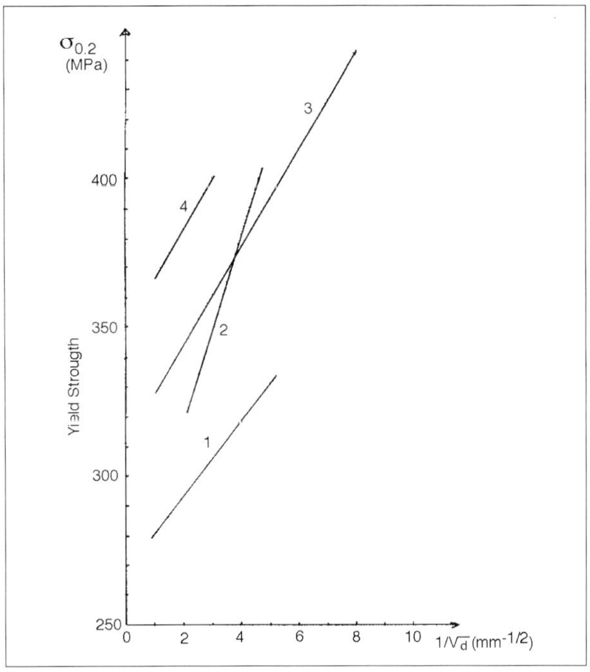

Figure 108 : Effect of Grain Size on Yield Strength

	C%	Mn%	N%	V%	
1	0.7	15.0	1.0		hot worked-W.Q.
2	0.7	15.0	1.0	1.0	hot worked-W.Q.
3	1.19	1,.2			hot worked-W.Q.
4	1.08-1.13	13.5-14.0		0-0.22	cast W.Q.

strength values, without changing the gradient. On the other hand an addition of vanadium markedly increases its angle coefficient κ. Although the testing of Petch's equation on one and the same material, generally gives more or less convergent results, the distribution of elements in solution in the austenite and the texture, may explain the differences noted between cast and rolled products (figure 108). The strange effect of nitrogen on yield strength of manganese austenites might be explained either by cluster formation or by nitrides, and that of vanadium by the formation of carbides. The addition of nitrogen to manganese austenitic steels, with the exception of non-magnetic steels used for retaining rings and steel used for valves, is not common, even though this element offers the possibility of obtaining a substantial increase in yield strength. Increasing the carbon content to 1.5% for a steel at 13% manganese, is

Base Alloy		Effect on Yield Stress $\sigma_{0.2}$ (MPa)									Ref.
C %	Mn %	C	Si	Mn	Ni	Al	Cr	Mo	V	Ti	
0.5	13	98.0	9.8	0	0		19.6	29.4	98.0		(142)
0.7	15 (1% Ni)	158.8					19.6	30.4	107.8		(143)
1.15	14.7				0		23.9	31.9			(74)
1.50	13.5					3					(27)
1.35	13.5								157.0	44.6	(144)

Table 19 : Effect of 1% Addition of Alloying Elements on Yield Stress $\sigma_{0.2}$

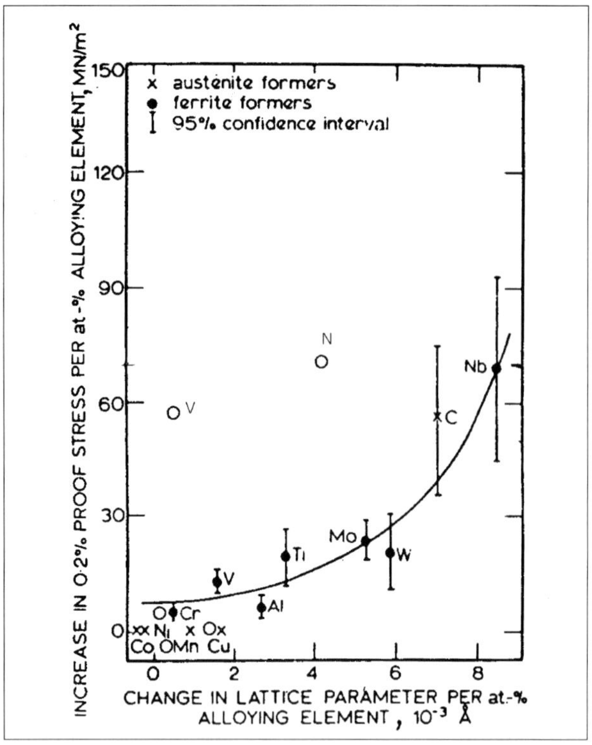

Figure 109 : Relationship between the Change in Lattice Parameter due to Solute Elements and the Increase in 0.2% Proof Stress (108)

beneficial if the section is no larger than 25 mm. For heavy sections, grain size increases, carbides precipitate during quenching, and carbon oversaturation of the austenite decreases, whence a lower value of yield strength and a marked drop in toughness. For example, the yield strength of a steel with 12% manganese and 1.5% carbon drops about 60 MPa, and elongations decrease by more than 20%, when the section changes from 25 to 150 mm (15).

It is not advisable to use the carbide-forming elements chromium, vanadium, and titanium because of their negative effect on toughness. An effect which is increased in large sections. Molybdenum, by slowing down carbide precipitation kinetics, is an exception. It provides an increase in yield strength without loss of toughness. The ASTM norm provides for the addition of molybdenum in two cases :

- to compensate for the loss of yield strength in grade 5 with less than 1.05% carbon, suggested for heavy section pieces,
- to prevent plastic deformation of the 1.10-1.20% carbon grade.

Chemical Composition	σ_B		$\sigma_{0.2}$		A%		KCU dJ/cm²			
	NWH	WH	NWH	WH	NWH	WH	NWH	WH	NWH	WH
C=1.10% Mn=18%	901	1348	421	1150	60.7	11.2	338	81	287	49
C=1.10% Mn=13%	751	1250	402	1160	30.0	5.5	290	52	136	21

Table 20 : Mechanical Properties of Austenitic Steels with 13 & 18% Mn in the Quenched State at 1100°C & in the Work-Hardened Stat (25% deformation)
NWH = Quenched State WH = Work-Hardened State

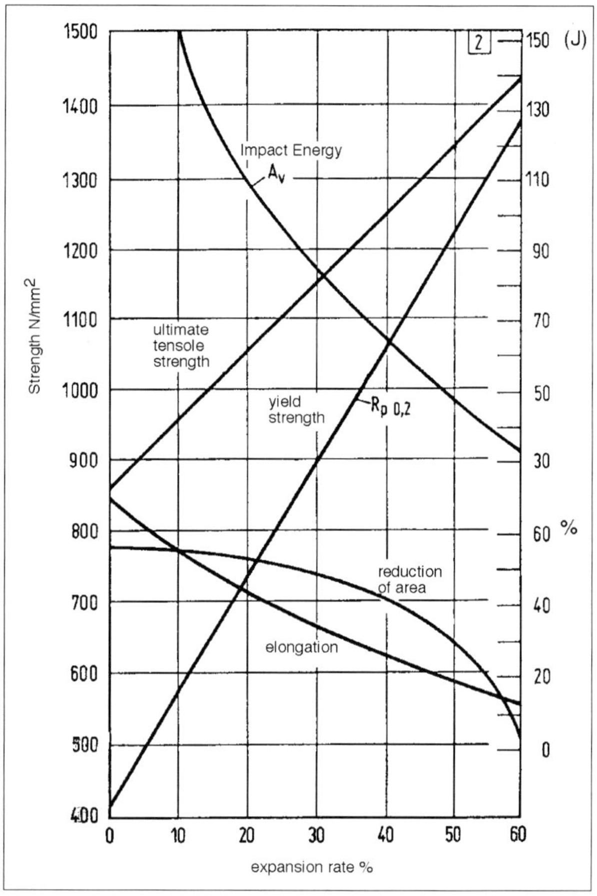

Figure 110 : Influence of Expansion Rate on the Mechanical Properties of Steel X55 Mn Cr 18 (145)

The possibilities provided by changing the chemical composition, to compensate for an insufficient yield strength, are limited, and do not always prevent unacceptable service deformation. One can have recourse to work-hardening, or to hardening through carbide precipitation, so long as the loss in toughness is compatible with the envisaged usage. Figure 110 shows the variations in mechanical properties of a retaining ring made of steel X55 MnCr(N) 18 K, according to the expansion rate applied. The anisotropy is very large, for a yield strength of the order of 1200 MPa. Elongations measured radially are of the order of 1/10 the elongations measured tangentially, and toughness is of the order of 1/5 (146).

The loss in toughness caused by work-hardening can be compensated to a certain extent by increasing the manganese content (table 20) or by adding austenite-stabilizing elements such as nickel and copper.

The hardening of manganese austenites through carbide precipitation, is accompanied by an increase in yield strength and a decrease in toughness. The different grades can be divided into two groups : forged grades (steel for retaining rings, and steel with 30 Mn - 10 Al - 1 Si - 1 C) which, for the same hardness, have a higher yield strength and greater elongations than cast grades (12 Mn - 2 Mo, and vanadium grades) (figure 111 a and b). This division is fully explained by the microstructure, the forged grades having no intergranular carbides, and a uniform distribution of vanadium carbides, or of phase κ $(FeMn)_3 AlC_x$, with a diameter of the order of 0.015 µm, whilst the cast grades present some intergranular carbides, and a larger precipitation of carbides, frequently located in the area surrounding the grain boundary.

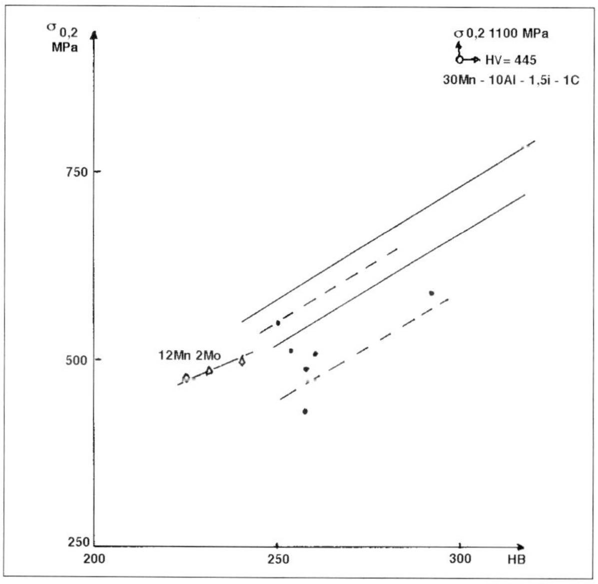

Figure 111a : Relationship between the Yield Strength $\sigma_{0.2}$ and the Hardness HB of a Manganese Austenitic Steel after Carbide Precipitation Treatment

The effect of section on mechanical properties obtained after carbide dispersion treatment has been reported only in the case of the steel with 12 Mn - 2 Mo (15). In 150 mm section, the steel 12 Mn - 2 Mo, compared with the steel of 12% Mn, has a higher yield strength at 70 to 100 MPa, and greater elongations (figure 112). In 150 mm section the level of toughness becomes insufficient above 1.2% C, for service reliability to be ensured.

b) Work-hardening

Work-hardening is evaluated with the aid of different criteria. The most often mentioned in technical literature is an increase of the stress, or the hardness, after a given deformation. Test conditions vary : deformation may be applied either by tension, or compression, or rolling, and the amount of deformation may vary from less than 1% to 50% of the initial state. Numerical data are generally not comparable from one author to the next. They only allow the qualitative effect of a few parameters on work-hardening to be shown for deformation rates, which have no relation to superficial deformations which come into play during sliding and abrasive wear.

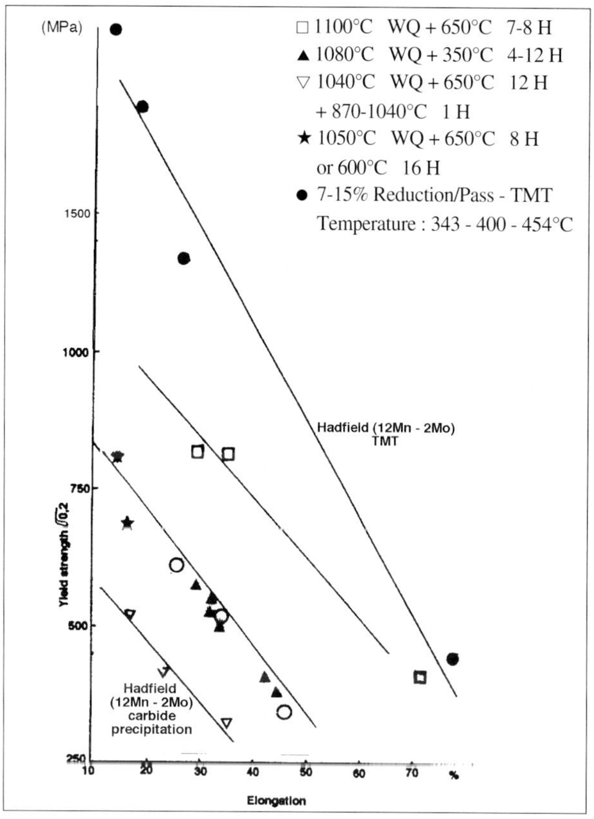

Figure 111b : Relationship between the Yield Strength $\sigma_{0.2}$ and Fracture Elongations of a Manganese Austenitic Steel after Carbide Precipitation Treatment and Thermo Mechanical Treatment (TMT)

	C	Mn	Cr	V	Ti	Ni	Mo	Ref
□	0.43	18	0.01-3.5	1,2				123
▲	1.33	12		0.3	0.18		0.75	125
▽	1.20	12					2	75
★	0.35 0.60	14 18		0.25 0.4		2 4	0 2.5	124
●	1.15	12.5	0.33				2.01	77
○	1.0 Mn	30 Al	(1Si-10Al)					56

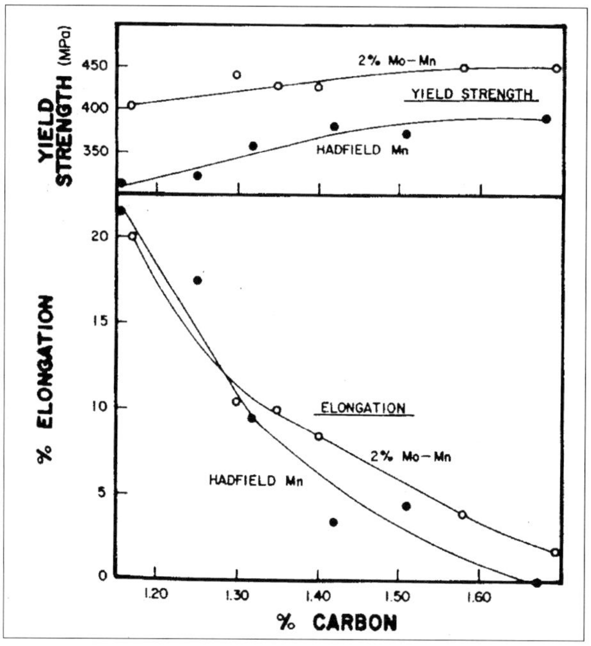

Figure 112 : Comparison of the Tensile Elongation and Yield Strength of Hadfield Manganese Steel and Dispersion-Hardened 2% Mo-Mn Steel in a 150 mm Section (15)

Work-hardenability increases with decrease in grain size (figure 113) (71) and also when the instability of the austenite increases, i.e., when manganese and carbon contents decrease (table 21). Alloying elements have little effect on the work-hardenability of stable austenites (figure 114). On the other hand, elements such as molybdenum, which increases the instability of the austenite, promote the work-hardenability of metastable austenites.

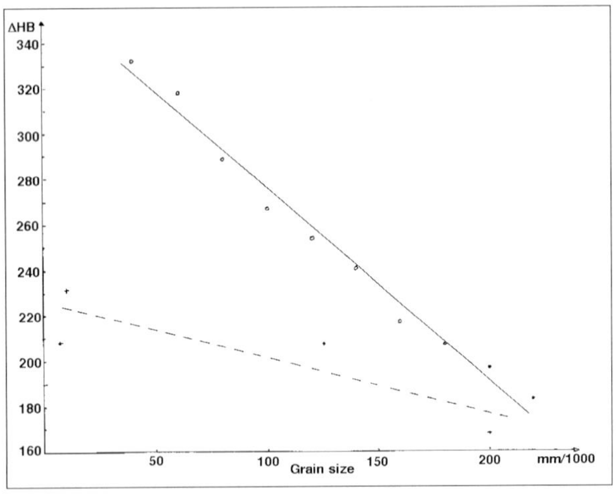

Figure 113 : Effect of Grain Size on Work-Hardenability (Increase in Hardness ΔHB after Deformation)
○——○ (Deformation unspecified) (147)
+ - - -+ (30% Deformation) (115)

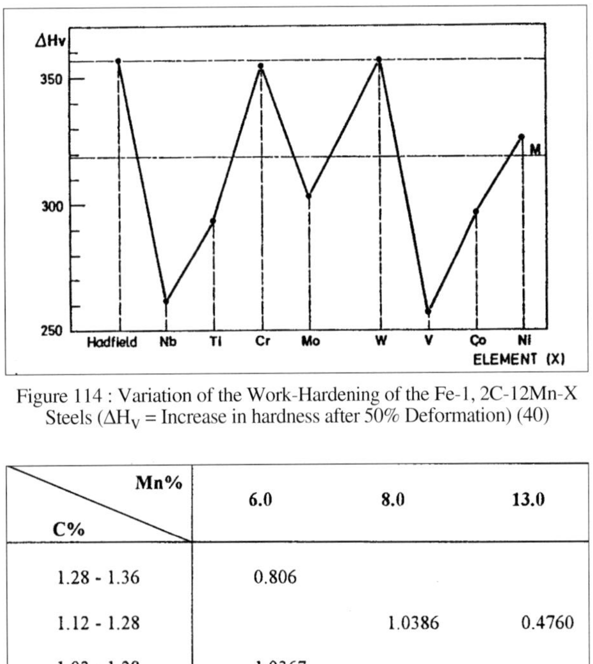

Figure 114 : Variation of the Work-Hardening of the Fe-1, 2C-12Mn-X Steels (ΔH_V = Increase in hardness after 50% Deformation) (40)

Mn% / C%	6.0	8.0	13.0
1.28 - 1.36	0.806		
1.12 - 1.28		1.0386	0.4760
1.03 - 1.28	1.0367		
1.03 - 1.08	1.087		
0.72 - 0.84		1.391	

Table 21 : Effect of C & Mn Content on Work-Hardenability of Mn Austenites (Increase in hardness ΔHRC in relation to % deformation ΔHRC/ε%) (125)

The addition of aluminium to austenites with 13% Mn, 1.0-1.25% C, has a negative effect on work-hardenability, but at higher carbon contents (1.30-1.75%) work-hardenability is improved, particularly between 150-300°C (148) (149).

c) Fracture Elongation

Fine-grained Hadfield steel, in the cast and quenched state, has higher tensile properties than coarse-grained (71) (65) (150). The elongations (ε%) decrease according to the following relationship :

$$\varepsilon\,\% = 44 - 6.61\,\frac{1}{\sqrt{d}}$$

where ε = elongation
d = grain diameter

established for a Hadfield steel cast at different temperatures so as to obtain grain diameters in the range 0.06 and 0.82 mm.

Fracture elongations of fully worked manganese austenites are practically independent of grain size. Forged and quenched Hadfield steel with 1.13% C and 13.5% Mn, and a grain diameter of 0.22 - 4 mm, and the non-magnetic steel 0.6C - 14Mn - 2Ni - 2Cr, rolled to different thicknesses (200 to 6 mm), with a grain diameter of 0.026 - 0.177 mm, both present variations in elongation of ±5% about the average (71) (110).

The effect of grain size on cast steel elongations depends on the solidification conditions, which govern both the grain size and the dendritic segregation. The surface/volume ratio of the grain and the dendritic segregation of the phosphorus change with an increase in grain size. Thermomechanical transformations homogenize the austenite, the phosphorus is more uniformly distributed, and its negative effect on the cohesive strength of the grain boundaries is neutralized, independentally, to a certain extent, of the nominal phosphorus content (figure 67).

The effect of manganese and carbon contents on fracture elongation is shown in figure 115 in the form of contour lines in the Mn-C plane. The composition range is extensive and concerns austenite-martensite steels, metastable austenitic steels of different groups and stable austenites, cast in thin sections.

The contour lines have a 'V-form', and as the V closes up, its cusp is displaced towards a higher manganese content and the carbon content remains around 1.10%, when the curve of equal elongation is displaced towards a higher value. The same elongations are obtained on the branches of the V, at a manganese content higher than the minimal content, with two different carbon contents.

Manganese markedly improves ductility by stabilizing the austenitic structure. The deformation

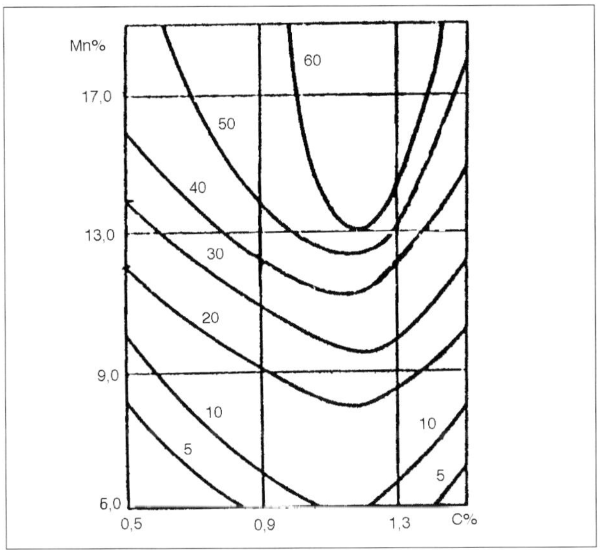

Figure 115 : Relationship between C and Mn Contents and Tensile Fracture Elongations (151)

rate needed to induce a martensitic transformation in metastable austenites, as well as the amount of carbides in austenites with high carbon content both increase with an increase in manganese content, and premature fracture, initiated by the martensite or the carbides, is delayed until more significant deformation takes place.

Commercial grades of manganese austenitic steels have elongations which tally, in thin sizes, with figure 115. An increase in section, which encourages dendritic segregation and slows the cooling rate during quenching, considerably reduces the mechanical properties. Thermomechanical transformations, which to a great extent neutralize the segregation effect, enable high mechanical property values to be retained, so long as the cooling rate during quenching is rapid enough to avoid any intergranular carbide precipitation. The Hadfield steel (1.1% C - 12% Mn) and non-magnetic steel (0.6% C - 14 Mn - 2 Ni - 2 Cr) retain their properties up to a size of the order of 125 to 250 mm.

The diffusion rate of phosphorus, silicon, and carbide-forming elements is too slow to remove the dendritic segregation of these elements by solution treatment of cast products. The effect of phosphorus is not neutralized as for rolled products, and the segregation of carbide-forming elements can amplify the formation of intergranular carbides during quenching.

Carbon content has an important effect : there is little variation in elongation up to 1.5% carbon in 25 mm section, but at 1.15% carbon, elongations fall from 45 to 20% when the section changes from 25 to 150 mm (figure 116). As with carbon, silicon does not change the mechanical properties up to a content of the order of 2.0%. Silicon content becomes critical from a certain section which depends on the chemical composition of the alloy : for example, elongations of

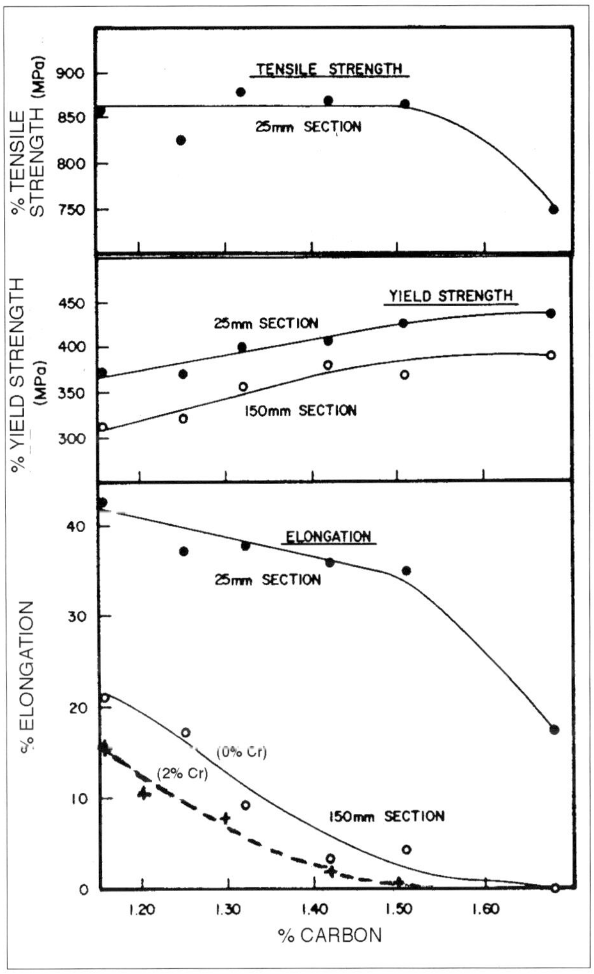

Figure 116 : Comparison of 25 mm and 150 mm Section Tensile Properties of Mn Steel (Base composition - 13% Mn, 0.6% Si, 0.035% P) (15)

a steel with 1.24% carbon, 13% manganese, and 0.035% phosphorus, in a 150 mm section, fall from 17 to 5% when the silicon content changes from 0.6 to 1.15% (15).

Figure 117 shows the effect of alloying elements on fracture elongation in Hadfield steel cast in thin sections. High contents of nickel slightly improve elongation : as nickel promotes carbide precipitation, it can be assumed that it will have a negative effect on heavy sections. Molybdenum has no effect up to 2%; at the same level chromium begins to initiate an appreciable decrease in elongation, which becomes accentuated in heavy sections. Elongations of a Hadfield steel with 1.10% carbon, in 150 mm section, fall from 22 to 15% after addition of 2% chromium (figure 116). Titanium, vanadium and rare earth elements considerably reduce elongations, except at low contents and in thin sections, where grain size refinement and the reduction of dendritic segregation, outweigh the negative effect of the added element (147) (152) (153) (154) (156).
The improvement disappears with an increase in section, and if the casting temperature is too high.

The combined addition of 0.005 to 0.5% vanadium and 0.008 to 0.02% boron, under controlled

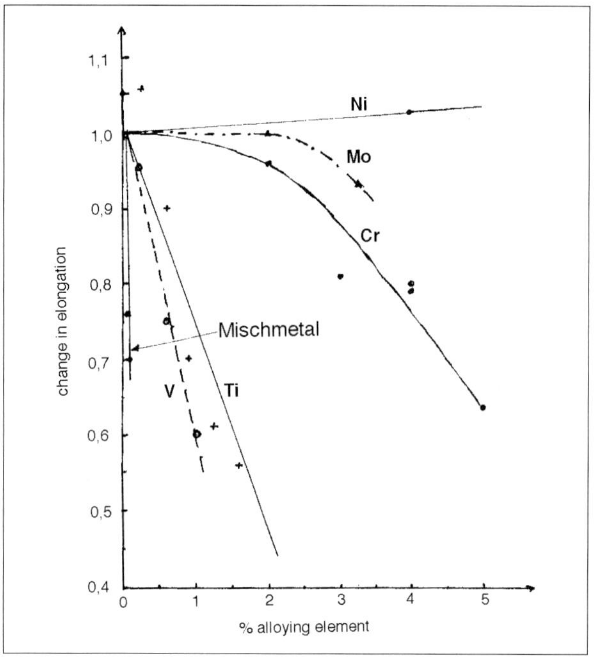

Figure 117 : Influence of Alloying Elements on Fracture Elongation
Cr-Ni-Mo (74), Cr-V (94), Ti (144), Mischmetal (156)

conditions, produces 60% elongation values, deep in the equiaxial zone, and 30% elongations in the surface columnar zone of a Hadfield steel of 120 mm section. The method of addition suggested is as follows : an aluminium deoxidation, to give a residual aluminium content of 0.02 to 0.09% is followed by adding titanium, up to 0.005%, to fix the residual nitrogen and thus avoid the formation of unwanted aluminium nitrides, and then proceeding with the addition of vanadium and boron (155).

d) Absorbed impact energy-temperature curve

According to A. Shulte and his colleagues (150), the absorbed impact energy, ductile to brittle, transition temperature of cast Hadfield steel decreases and the impact energy at room temperature increases, with a decrease in grain size (figure 118). Wlasow and Kosudora (157) confirm the favourable effect of fine grain on room temperature impact energy (figure 119). On the other hand, the results of H. Berns (71),

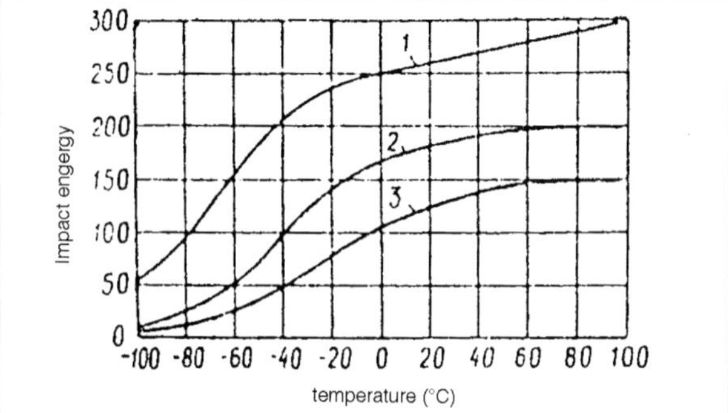

Figure 118 : Influence of Casing Temperature (Grain Size) on Absorbed Impact Energy-Temperature Curve of a Hadfield Steel (150)
b. Casting temperature 1) 1385°C, 2)1440°C, 3) 1470°C ↓coarser grain

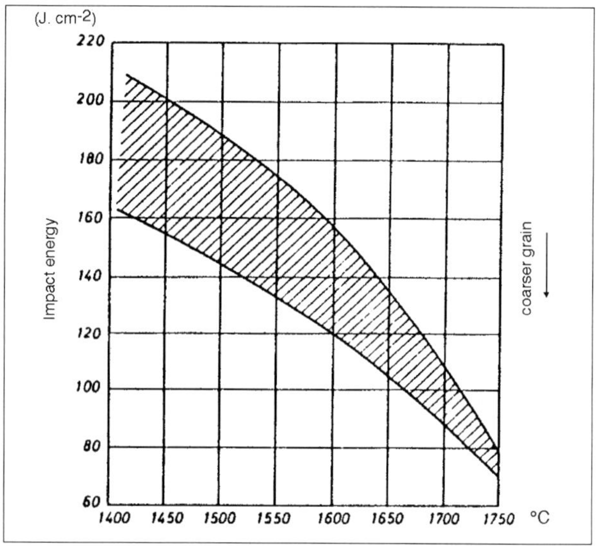

Figure 119 : Influence of Casting Temperature (Grain Size) on Impact Strength at Room Temperature of a Hadfield Steel (157)

obtained with test pieces taken from the equiaxial solidification zone of Hadfield steels, show that the impact energy transition temperature of a coarse-grained steel is lower, and the room temperature impact energy is higher, than that of a fine-grained steel (figure 120). The latter point is supported by the results of Z. Krolikowski and W. Ferns (158) (figure 121).

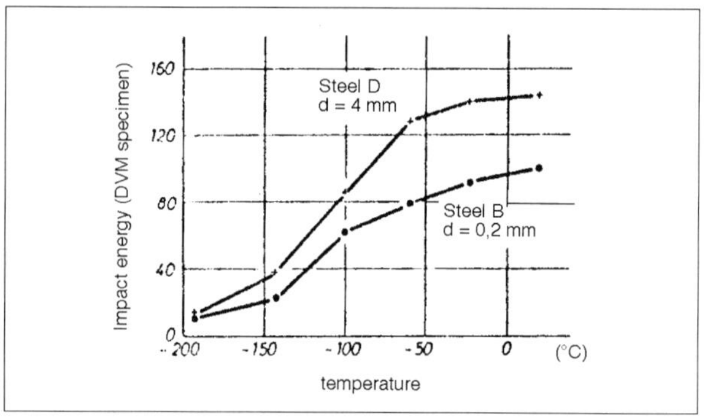

Figure 120 : Influence of Grain Size on Impact Energy Transition Temperature (71) (DVM specimen)

Steel	C%	Mn%	Casting T°
B	1.12	13.5	1430°C
D	1.15	13.6	1460°C

4 hrs - 600°C + 30 min - 1070°C WQ Equiaxial Structure (71)

The contradiction of experimental results concerning the study of the relation between grain size, impact transition temperature and the value of impact shelf energy, could be explained by the embrittlement of the fine grain steel.
The effect of manganese and carbon contents on impact energy at room temperature of thin section cast austenites is given in figure 122 in the form of contour-lines, which traverse the Mn-C plane in different austenite stability domains. The contour lines under consideration are similar to contour lines of equal elongation given in figure 115, and they show that an increase in manganese content improves both impact energy and elongation.

The impact energy transition temperature of the steel with 12% manganese, is practically independent of

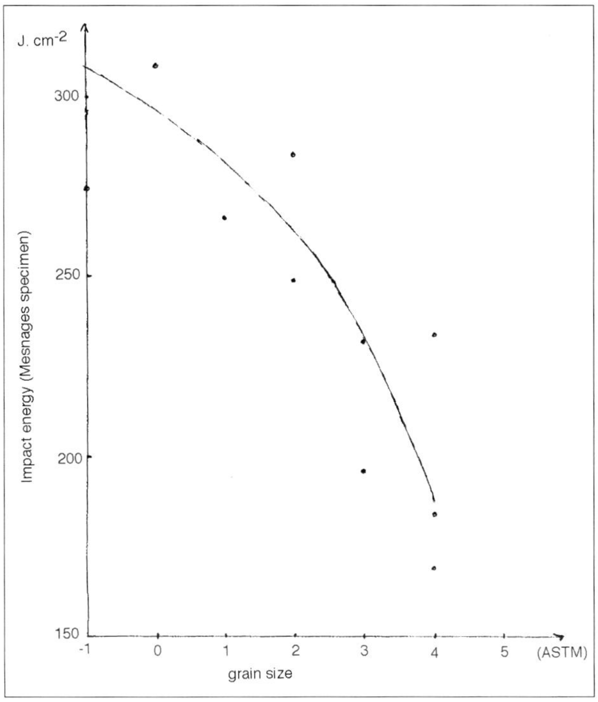

Figure 121 : Influence of Grain Size on Absorbed Impact Energy at Room Temperature (158) (Hadfield Steel C = 1.22%, Mn = 12.20%, Si = 0.66%, P = 0.036%, S = 0.01%, Cr = 0.18%, Ni = 1.04%) Quenched at 1080°C

carbon contents between 0.7 and 1.4%. The impact transition temperature of the steel with 0.97% carbon, diminishes rapidly when manganese content increases from 8 to 10%, and then tends towards an asymptotic value at around 14% (figure 123).

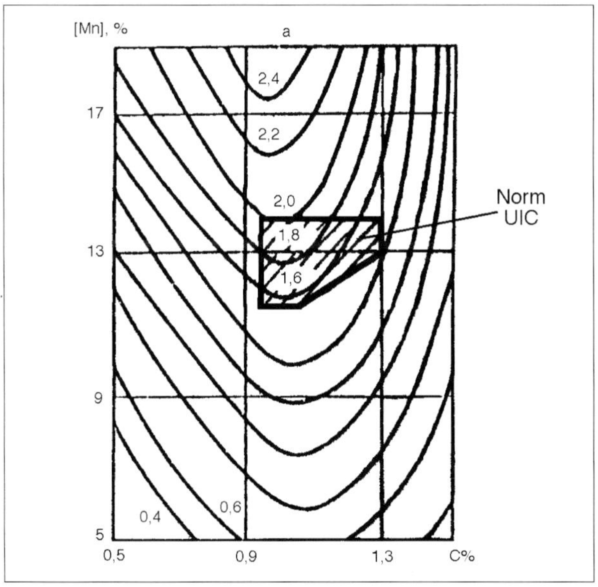

Figure 122 : Relationship between C and Mn Contents and Absorbed Impact Energy Mesnager Specimen (10^2.J.cm^{-2}) (151)

The carbide-forming elements chromium, vanadium, titanium and boron, lower the impact shelf energy and raise the impact transition temperature. The impact shelf energy decreases in proportion to the concentration in carbide-forming elements, which, in a decreasing order of harmfulness, are as follows : Ti, B, V, Cr. The carbide-forming elements rapidly raise the impact transition temperature which then seems

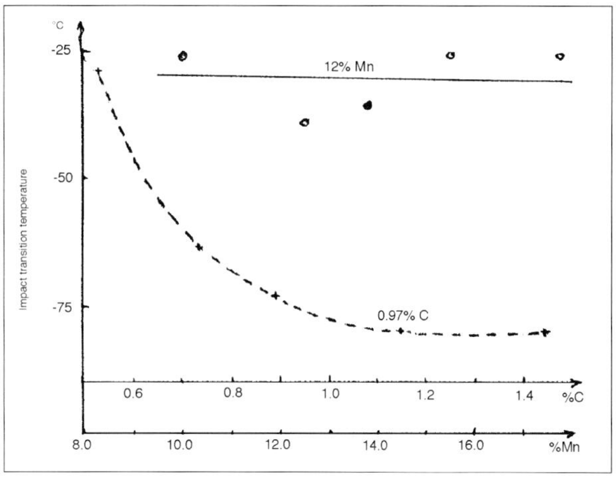

Figure 123 : Influence of C and Mn Content on Impact Strength Transition Temperature of Manganese Austenitic Steels Cast in thin Section (159)
●——● Influence of Carbon (12.2% Mn)
+ – – –+ Influence of Manganese (0.97% C)

to tend towards an asymptotic value. Ranged in the order of harmfulness, these elements have a somewhat different lineup : B, V, Ti, Cr (figure 124). The cross-sectional dimensions of castings brings two parameters into play : dendritic segregation, and transformation kinetics of the austenite in continuous

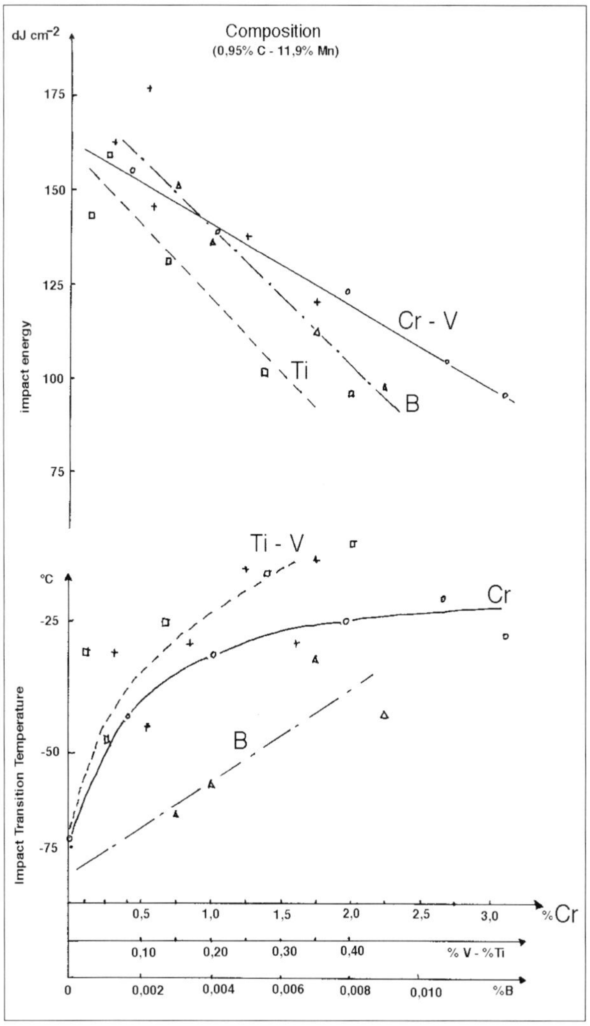

Figure 124 : Influence of Cr V and B on Impact Transition Temperature and absorbed Impact Shelf Energy of an Austenitic Steel with 0.95% C and 11.9% Mn (159)

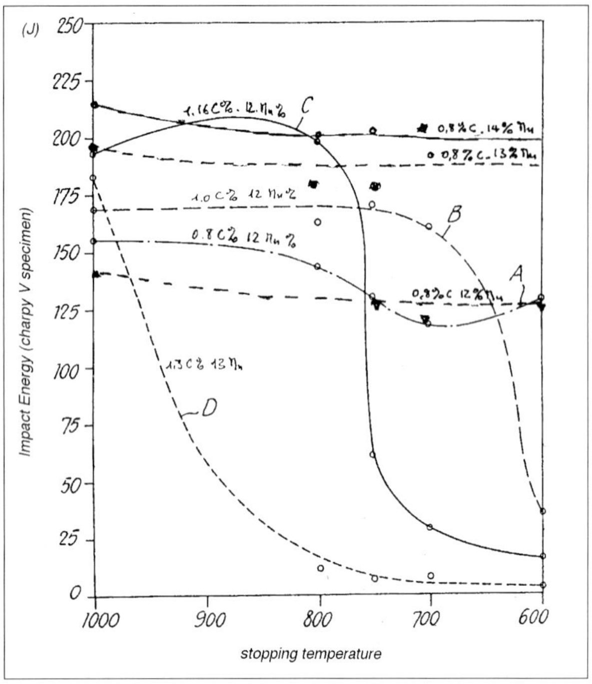

Figure 125 : Influence of C and Mn Content on the Embrittlement Domain during Cooling (160) (Cooling from 1000°C to Stopping Temperature in 30 min and WQ)

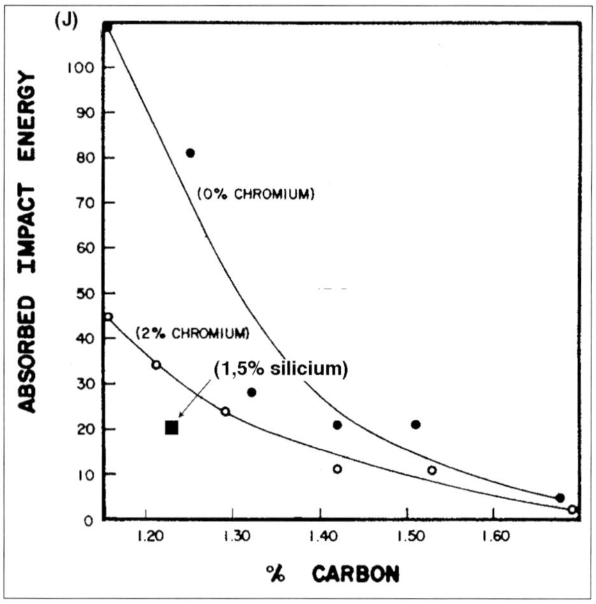

Figure 126 : Effect of Carbon and Chromium on the Izod Absorbed Impact Energy of 150 mm Section Mn Steel. Base Composition - 13% Mn, 0.6% Si, 0.035% P) (15)

cooling, which govern the embrittling factors. Embrittlement susceptibility during quenching has been evaluated by measuring the impact energy, after sudden interruption of the quenching cycle at 1000 and 600°C by drastic cooling. Using this method, study of austenitic steels with different carbon and manganese contents shows (figure 125) that steel with 12% Mn and 0.8% C is not embrittled under experimental conditions, whilst steels with higher carbon contents become embrittled from a temperature which is higher as the carbon content increases. At a carbon content of 0.8% impact energy increases, when the manganese content increases, and remains unaffected by embrittling conditions.

As it is difficult to simulate the global effect of the parameters which play a role in the mass effect, tests can only be carried out on samples taken from test pieces made from large commercial cast parts. There is little data available, but figure 126 shows the effect of carbon content on the Izod impact energy of a steel with 13% Mn, cast in 150 mm section, the impact energy decreasing very rapidly when the carbon content increases. An addition of 2% Cr or of 1.5% Si, considerably lowers the impact energy (figure 126).

IV. Contact Deformation

In most applications of Hadfield steel and of metastable austenites, (wear parts, and railway crossings), the material is submitted to contact points with its counter-check, at high stresses, which produce superficial deformation. The surface hardness of layers work-hardened in this way is considerably higher (500 to 900 Hv) than that reached by tension after fracture (Hv ≃ 450). The superficial work-hardening observed, results not only from a high work-hardening propensity, but also from Hadfield steel's extraordinary capability for deformation when the deformation modes produce a gradient of deformation.

a) Superficial-hardening

1. Surface work-hardening by shot peening or explosion

Surface hardness and the hardness gradient of layers work-hardened by shot peening, increase with the instability of the austenite (figure 127). Surface deformation depends on shock parameters. The depth

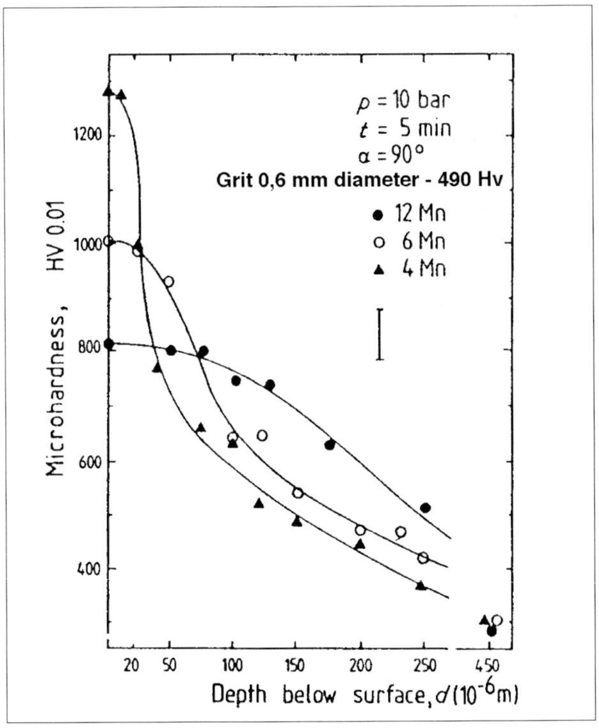

Figure 127 : Plot of Microhardness Against Depth Below Shot-Peened Surface (61)

which is work-hardened is proportional to the diameter of the shot peening and increases with the rate of impact (162). Consequently, the curves of figure 127 are representative only in the operating conditions described.

The work-hardened layer of stable austenites presents a very great density of micro-twins, which decreases in depth towards undeformed metal. The work-hardened layer of metastable austenites is made up of an ultra-thin surface layer, more or less rich in martensite, and of a sub-layer deformed by slip. The nucleation of the martensite occurs at the interface and at the intersection of the slip bands. The thickness of the martensitic layer increases with the instability of the austenite and the rate of impact. Martensite saturation is reached after a short period of shot peening (figure 128).

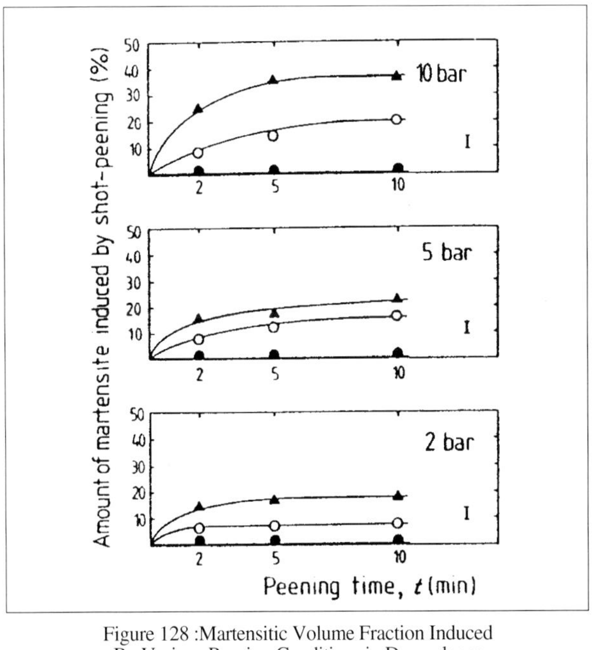

Figure 128 :Martensitic Volume Fraction Induced By Various Peening Conditions in Dependence on Peening Time (161)
▲ 4 Mn - 1.15 C ○ 6 Mn ● 12 Mn

Major deformation of component parts, and damage to the work surface, are the two main problems posed to equipment subject to repeated impacts and can eventually be avoided by proceeding beforehand with a work-hardening comparable to that measured in service. Shot peening has too superficial an effect to bring about improvement. In applications where the service work-hardened layer is thicker, work-hardening by hammering is used, or, more often, by explosion, the latter process having the advantage of not producing any distortion. By using penta-erythritol tetranitrate in plastic sheets, a superficial hardness can be obtained for Hadfield steel of the order of 310 Hv with an explosive shock loading of 0.3 gCm^{-2}, and of the order of 580 Hv with an explosive shock loading of 0.6 gcm^{-2}. The work-hardened layer extends to a depth of about 35 mm (figure 129) (163) (164) (165).

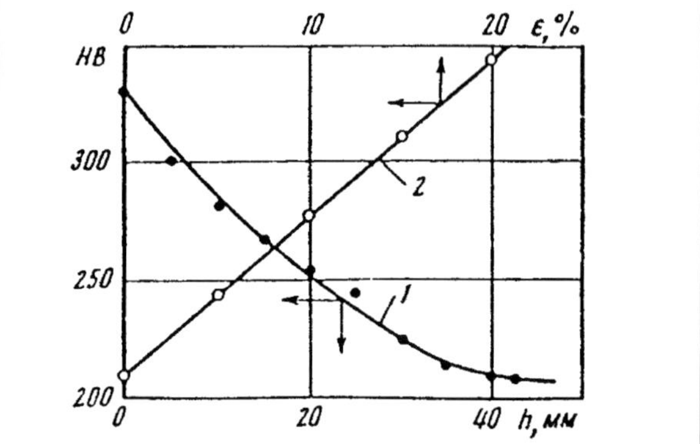

Figure 129 : 1. Hardening by Explosion - Hardness-Thickness Curve
2. Hardening by Compression - Hardness-Deformation Curve (164)
(1.25% C - 13.8% Mn - 0.68% Si)

Explosive hardening of railway crossings made of Hadfield steel, is regularly practised in certain countries (the USA for example). The surface hardening of worn railway crossings is of the order of 530 to 580 Hv. Experimental study of the hardening of Hadfield steel on a device simulating the railroad system, shows a hardness saturation of around 500 Hv from about 50 impacts (117). Explosive hardening produces hardness levels comparable to those obtained by service hardening. Authors have suggested using dispersed carbides in Hadfield steels, to avoid explosive hardening (117) (126).

2) Superficial Work Hardening by Repeated Impact

Change in the residual stress state is a preliminary of damage to the surface layers of Hadfield steel which has undergone repeated impacts. Figure 130 shows the evolution of surface stresses provoked by repeated impacts, of a quenched Hadfield steel work-hardened by explosion or shot peening. Before testing, the steel work-hardened by explosion is under biaxial tension and presents surface microcracks, whilst the shot-peened steel is in biaxial compression. The tensile stresses of steel hardened by explosion decrease rapidly after a few shocks. The behaviour of quenched and shot-peened steels is different : the direction of the stresses is reversed, and they increase rapidly. The dissimilarity in stress state change can be attributed to the difference in thickness

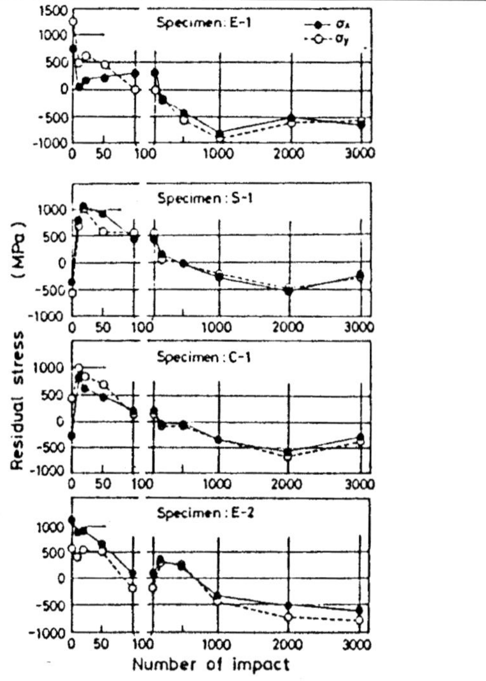

Figure 130 : Change in Surface Residual Stress due to Repeated Impact Loads (166)

		Impact Energy	σ (Mpa)
E1	Explosion hardening	148.4 J	632
S1	Shot peening	148.4 J	632
C1	No working	148.4 J	632
E2	Explosion hardening	74.9 J	448

(Cast Hadfield Steel SC Mn H3 1050°C WQ)

of the work-hardened layer. Tensile stresses decrease after about 20 shocks, and disappear between 100 and 500 shocks. Afterwards, when the number of shocks increases a compressive state is reached, and after 2000 shocks the stresses are of the order of 500 MPa whatever the initial state (166).

Surface compressive stresses are balanced in the sub-layer by tensile stresses. The combined effect of internal stresses and applied contact stresses is the cause of different processes of surface damage, such as spalling, microspalling, flaking, and fracture.

A method of evaluating the high stress of grinding balls, provides a good simulation of these modes of damage. The test machine is made up of a torus sector with a capacity of 18 balls, and of a conveyor which raises the balls to a fall height of 3.5 m. A series of 22 balls, 75 mm in diameter, circulates in a closed circuit in the machine, a ball in free fall hits another situated at one end of the torus sector, the balls in contact in the container successively transmit the unabsorbed energy, and the last ball receives just the energy necessary for it to be ejected onto the conveyor to begin the cycle again.

Microspalling is the only cause of damage observed for a Hadfield steel. This appears after 100 to 150,000 cycles, in the form of oval or crescent pits with a diameter of 2 to 4 mm, to a depth of 2 to 3 mm. They would be initiated on the surface through fatigue, propagate from the surface to a zone of weak compression stresses, and finally the root would be fractured.

Balls of metastable austenitic steel are destroyed by sudden fracture after 1000 to 50,000 cycles, sometimes only a few hours after testing. The fracture generally occurs on an equatorial plane, and the thickness of the work-hardened layer is of the order of 5 mm. Fracture is due to stresses engendered by work-hardening and to an increase in volume of the deformed layer due to the induced martensite. The stresses of superficial compression are very high, and fracture occurs when the constituent of the internal tension stresses, which balances them, exceeds the fracture stress. The accumulated energy is considerable, and the fracture is almost explosive.

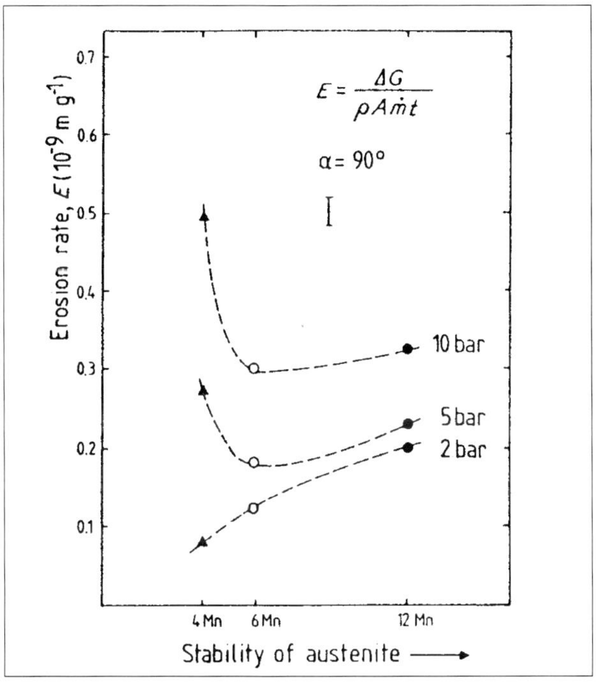

Figure 132 : Plot of erosion Rates Mesured at Various Peening Conditions against Stability of Austenite, i.e; wt%Mn (161)

In figure 131 impact wear damage resistance of Hadfield steel and metastable austenites is compared to that of other materials used for their wear resistance. Hadfield steel is ranked amongst the best performers.

Metastable austenites resist erosion better than Hadfield steel, insofar as the impact energy does not develop an excessive induced-martensite volume fraction (figure 132).

b) Sliding and rolling

1. Sliding

The hardness-depth curves of the layer work-hardened by unlubricated sliding are similar to the curves for the hardened layers produced by shot peening. The instability of the austenite increases the surface hardness and reduces the thickness of the deformed layer (figure 133). The wear resistance of austenite, under low load (110 N) is proportional to the surface hardness, the latter increasing markedly after cooling in liquid nitrogen, and hence wear resistance is significantly improved.

Wear resistance decreases when the applied load increases, the metastable austenites registering a drop in wear resistance which is all the greater the more they are unstable, and the wear-resistance/surface hardness relation is no longer linear (figure 134).

The microscopic structure of the layers hardened by sliding are similar to those hardened by shot peening. Stable austenites present, on the surface, a high density of micro-twins, which decreases with depth. The metastable austenite layer is made up of a very thin surface martensite layer, and of a sublayer

Specimen	*Description*	*Brinell hardness (HB 3000)*	*Number of balls*
Cast irons			
24	3C low alloy	667	3
41	9Cr Nin Ni hand IV	622	1
42	15 Cr 3Mo	540	2
19	20Cr 2Mo	680	3
43	20Cr 2Mo 1Cu, as cast	578	1
44	25Cr HC 250	648	2
Cast steels			
25	low alloy, 125	375	2
36	low alloy, AR 360	370	1
37	2Cr 0.5Mo	780	3
38	6Cr 1Mo	643	2
Manganese austenitic steels			
40	6Mn 1Mo, cast	595	3
30	6Mn 2Cr 1Mo, cast	461	3
34	12Mn, cast	200	1
56	12Mn, lorged	226	1
Wrought steels			
51	commercial ball A	712	2
9	commercial ball B	740	14
52	commercial ball C	627	2
76	commercial ball C:	321	4
57	commercial ball F	352	1
58	commercial ball I	670	8
55	univ. Calif. "Quatough"	411	2
23	ALRC Cr Mo	627	3

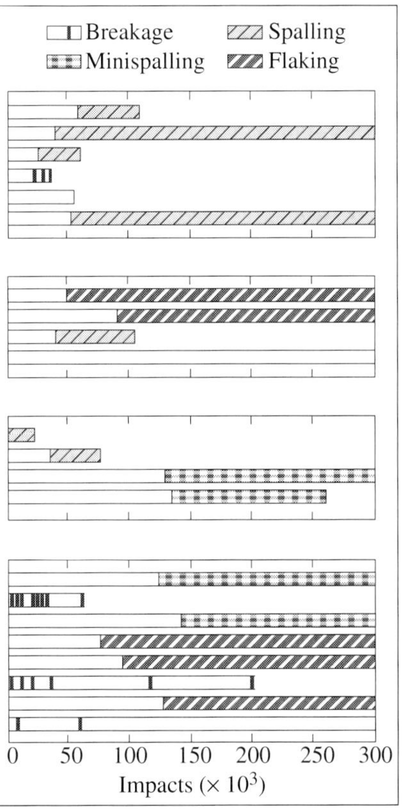

Figure 131 : Comparison of Repeated Impact Resistance of Hadfield Steel and Metastable Austenites with that of other Materials (167)

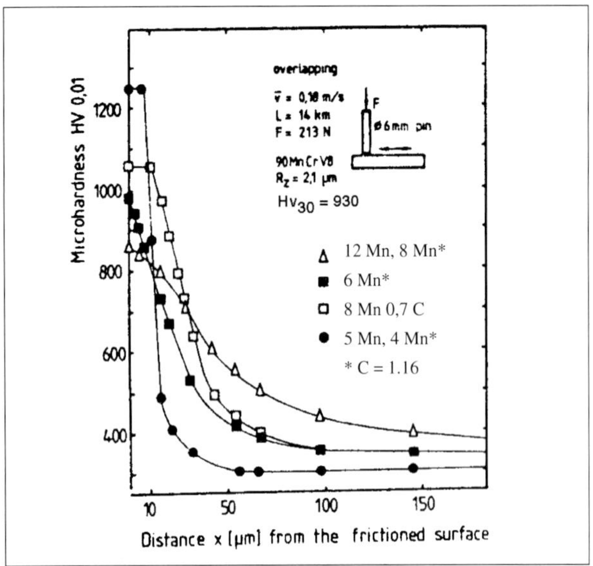

Figure 133 : Plot of microhardness HV0.01 versus Distance x from the Frictioned Surface. In Unaffected Depth the Microhardness of all Alloys Investigated goes down to about HV0.01 (11)

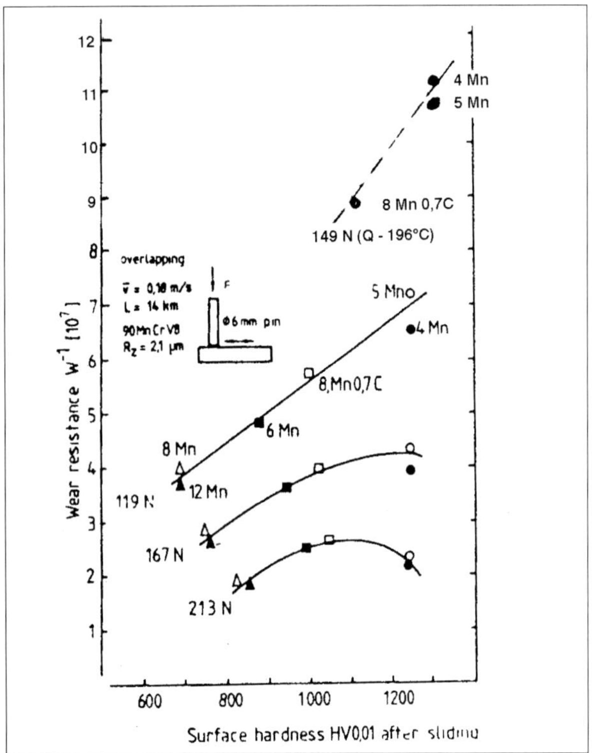

Figure 134 : Influence of the Applied Load on Wear Resistance and Surface hardness after Sliding (11)

deformed by sliding, where the nucleation sites of the martensite are visible at the interface and at the intersection of the slip bands. The thickness of the surface martensite layer increases with increase in the applied stress.

Examination of the rubbing surfaces with a scanning electron microscope shows that wear under low load proceeds from a mechanism of adhesive wear, with a transfer of the austenitic material to its counter-check. The surface layer cracks, and fragments break off, by a brittle failure mechanism, under heavier loads. The fragments behave like an abrasive, causing scratches by micro-cutting during their ejection (11).

Shot peening has a beneficial effect on the sliding wear resistance of Hadfield steel. The gain for metastable austenites is negligible, but shot peening could prove advantageous in less severe, rubbing conditions (figure 135).

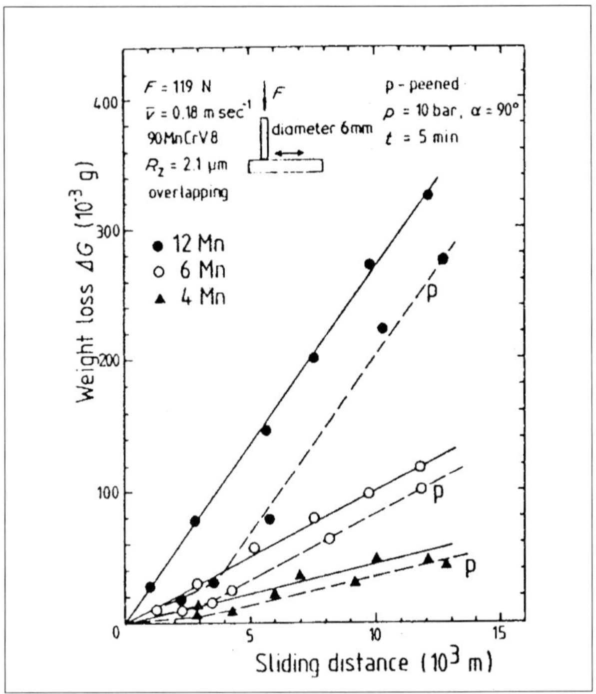

Figure 135 : Comparison of the Sliding Wear Behaviour of Peened (p) and Unpeened Specimen (pin-test apparatus) (161)

2. Rolling

Repeated contact stresses during rolling harden the surface layers, the process reaching saturation well before the fatigue limit, at 2.106 cycles (169). The profile of the hardness curve of the rolling-hardened layer is different from that of layers hardened by shot peening, sliding and abrasion wear. Hardness reaches a maximal value beneath the rolling surface, where it reaches a much lower value than for other surface deformation modes (figure 136a).

The maximum hardness of the work-hardened layer increases, according to a linear relation, when the applied contact stress increases. The hardness is higher in the absence of chromium or nickel (figure 136b). The distance of the maximum hardness with respect to the rolling surface is situated between those of the stress levels $Z_0 = 0.787b$ and $Z_1 = 0.50b$ (b = width of the contact surface) (figure 136c).

A fatigue crack is initiated at the level of maximum shearing stress, and begins to propagate in circular arcs towards the interior, then changes direction and propagates parallel to the surface, but at a greater distance than the maximum shearing stress.

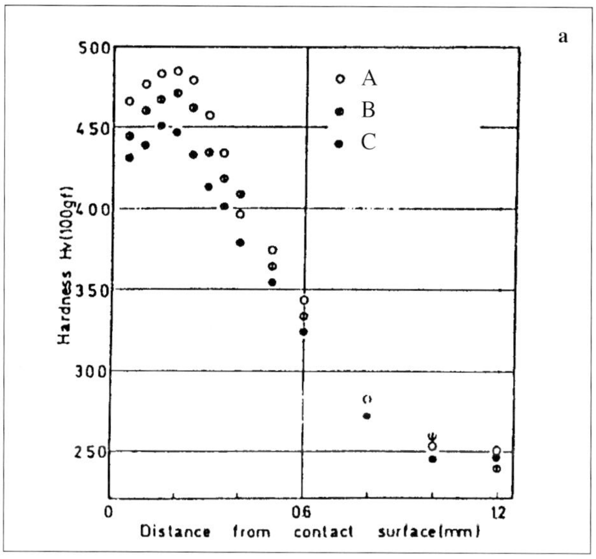

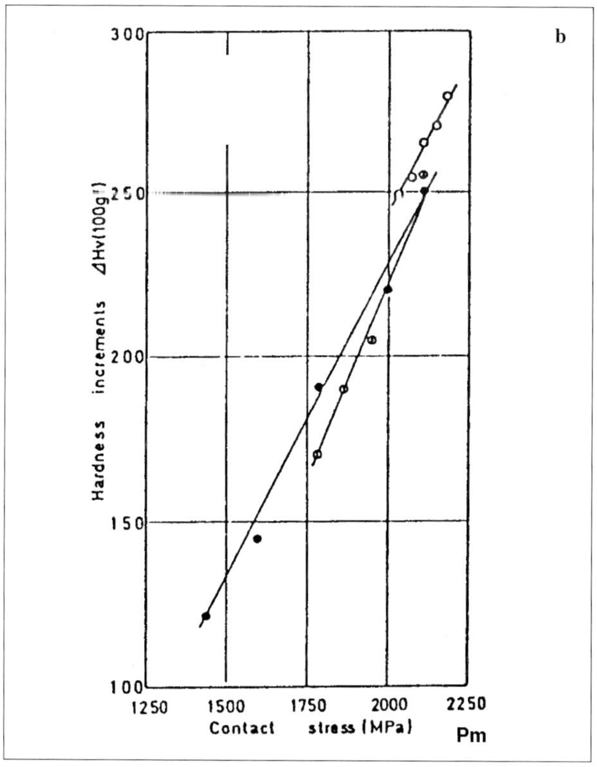

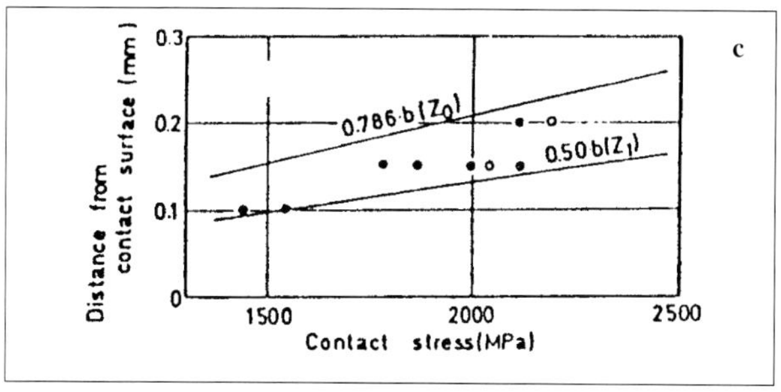

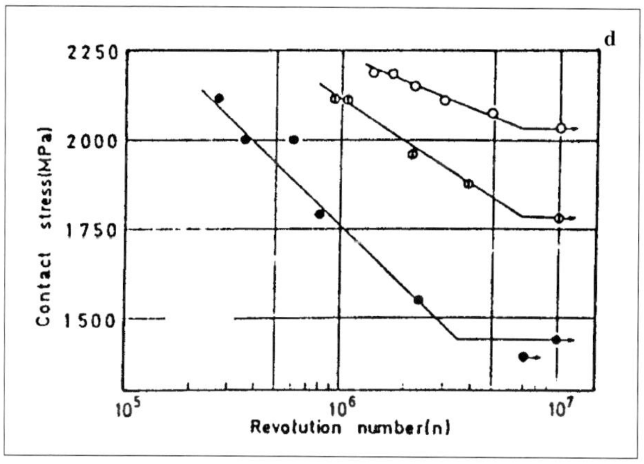

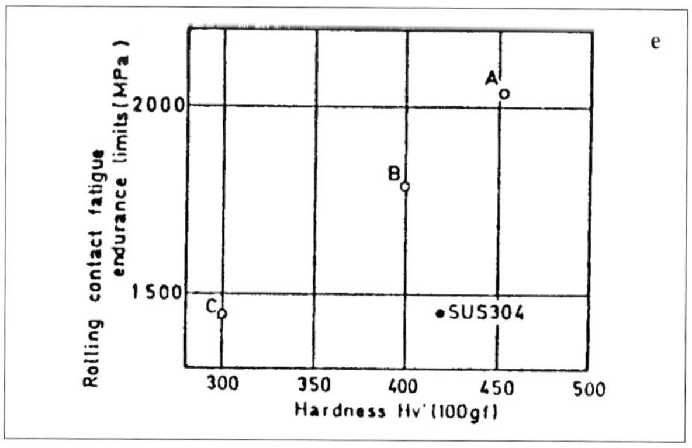

Figure 136 a,b,c,d,e,f : Rolling Fatigue Resistance of Manganese Austenitic Steels (168)

$$Pn = \sqrt{\frac{m^2EP}{\pi(m^2 - 1)BR}}$$

P. Applied change, B. Test piece width, R. Test Piece Radius, E. Young Modulus, 1/m Poisson increment

Test conditions : V = 800 rpm^{-1}, relative sliding -9% (1.26 ms^{-1}), lubrifiant - oil, test until 10^7 cycles

(result do not differ from those obtained in true rolling)

	C%	Mn%	Si%	Cr%	Ni%	σs(MPa)	HV0.1
A	1.05	12.33	0.62			392	204
B	1.04	13.34	0.74	2.06		431	230
C	1.04	12.83	0.73		3.83	353	180

The fatigue limit is reached after 10^7 cycles. Hadfield steel, without the addition of any alloying elements, has a higher fatigue limit (2035 MPa) than steel with 2% chromium (1785 MPa) and steel with 3% nickel (1439 MPa) (figure 136d). The fatigue limit of Hadfield steels is proportional to surface hardness. The stainless steel grade AISI 304 has a considerably lower fatigue limit, for comparable surface hardness, than manganese austenitic steels (figure 136e).
Wear from rolling and from non-lubricated rolling/sliding of Hadfield steel on a normalized steel with 0.7% C, 0.7% Mn, increases when the contact pressure increases. Explosive hardening of Hadfield steel improves its rolling wear resistance, but the difference is small for rolling/sliding (figure 137). Wear is greatly accelerated by pre-existing defects, caused by explosive hardening.

The worn surface of Hadfield steel is rough with pitting observed, whether it was work-hardened or not beforehand. The surface of carbon steel presented spalling. Wear fragments from rubbing/sliding were made up of flakes, showing micro-cracks and particles transferred from Hadfield steel.

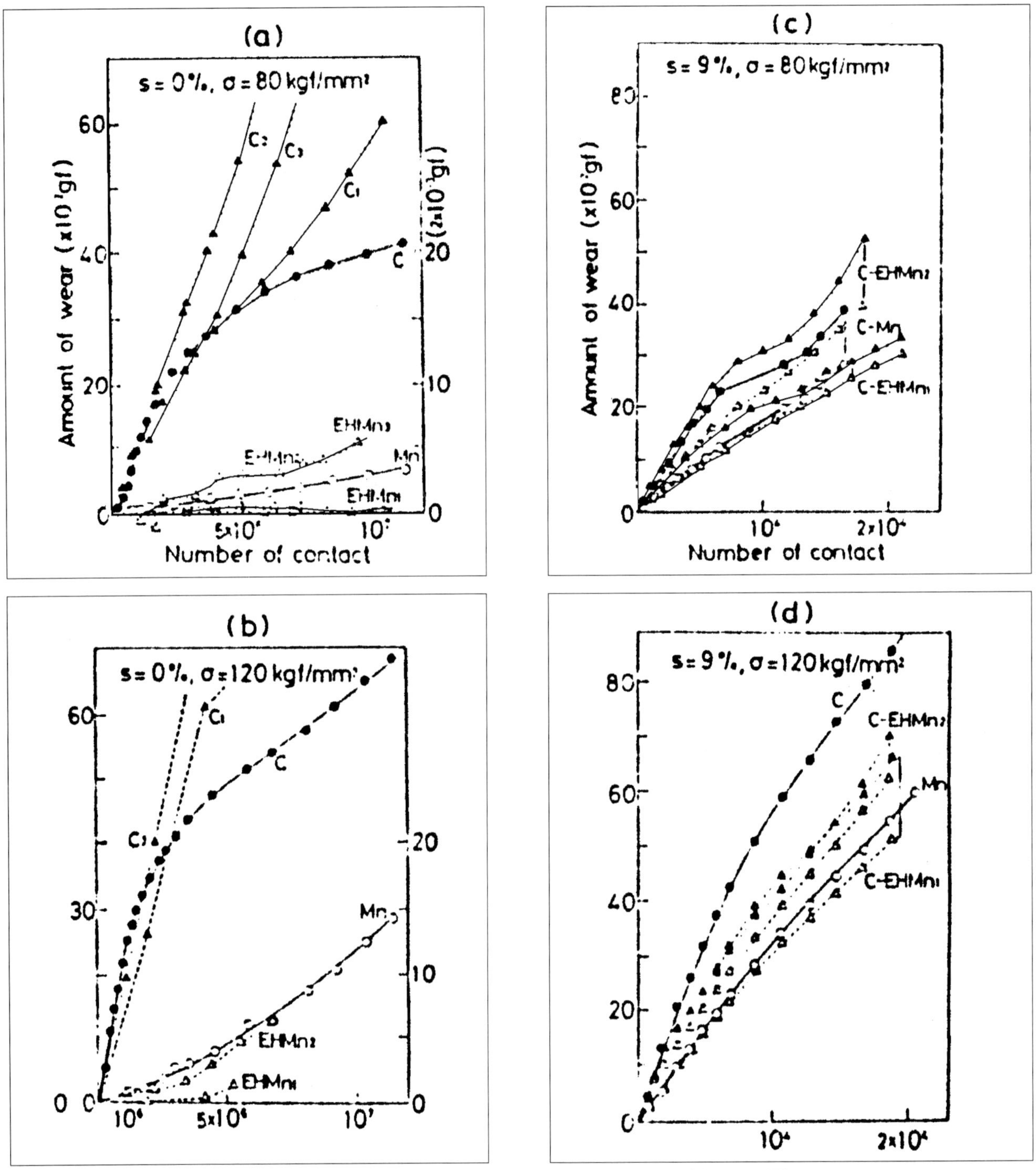

Figure 137 : Relationship between Amounts of Wear and Number of Contacts (170)
a & b = Rolling Wear - c & d = Sliding-rolling Wear

c) Abrasive Wear Resistance

Abrasion resistance develops from an indentation of the abrasive under the effect of a force applied at right angles and from its lateral displacement under the effect of a tangential force. This causes scratching through a cutting effect or a groove made by plastic deformation following its angle of attack. Wear corresponds to the sum of the volume of matter ejected in the form of microchips and of the volume of matter strongly work-hardened, which is eliminated by a brittle mechanism, by microspalling and in the form of secondary microchips. All configurations of abrasive wear, wear of two bodies, wear of three bodies, and erosion, are connected to this process.

The work-hardening of the material abraded, its hardness, mechanical strength, angularity of the abrasive, and the intensity of the applied loads, are the most important variables governing abrasive wear. Surface work-hardening, resulting from the indentation of abrasive particles and to the grooves they have traced, very rapidly modifies the initial contact condition. An equilibrium is then established between wear rate and the surface work-hardening. Figure 138 shows the effect of austenite stability, and of abrasive type, on the work hardening of the surface layer. The surface hardness of a metastable steel with 6% manganese, abraded by silicon carbide,

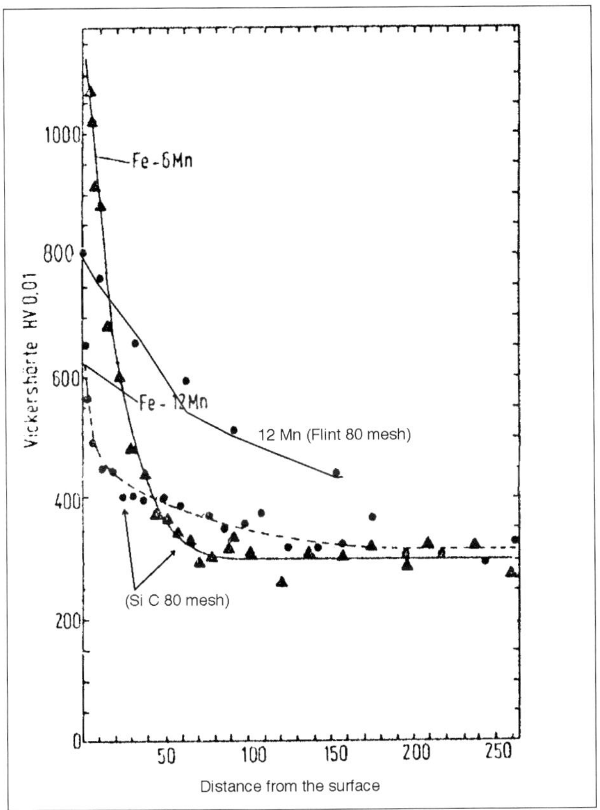

Figure 138 : Influence of Hardness of Abrasive, and Austenite Stability, on surface Work-Hardening (171)

is much higher than that of Hadfield steel : the difference is explained by the martensite volume fraction of the abraded surface. The presence of martensite on the abraded surface of Hadfield steel, shown by X-ray diffraction, is confirmed by the ferro-magnetic state of the microchips (173).

Flint, with similar particle size but lower hardness than silicon carbide, can work-harden Hadfield steel more, at the same apparent pressure, than silicon carbide. The deformed layer is thicker and the surface hardness greater. Flint's sharp edges would blunt more rapidly than those of silicon carbide and would lose their cutting ability, whence a more severe work-hardening, and a change in the relative ratio of the modes of abrasive wear through microchip formation and by brittle deformation.

The relative importance of surface work-hardening or abrasion resistance can be correlated with the elasto-plastic abrasive wear model proposed by Torrance (174) : according to this model, the abrasion resistance of pure metals and of metallic alloys obeys a linear relation whereby surface hardness plays the dominant role (figure 139). Hence the surface micro-hardness of abrasion resistant components' wear in the field is a good criterion for selecting abrasion resistant alloys, or for checking the abrasion conditions.
The tonnage of austenitic manganese steels used for their abrasion resistance properties is to a large part

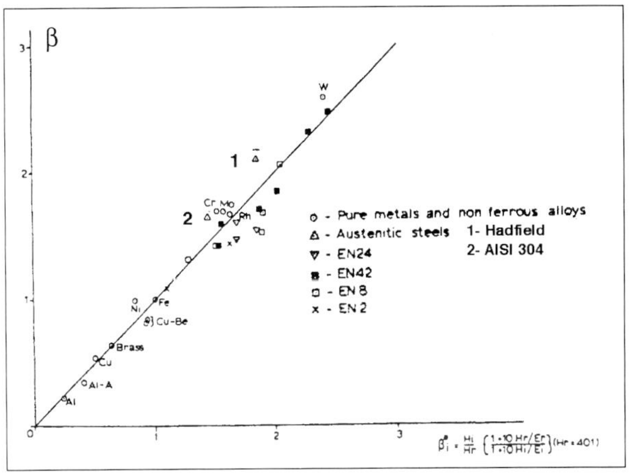

Figure 139 : βi vs βi* of Elasto-plastic Model of Abrasive Wear of Torrens (all the materials are situated on same curve) (174)

consumed in mine and quarry grinding facilities. Abrasive wear depends, amongst other things, on the components and on the intensity of the contact forces exerted in machines. Avery (176) places the contact conditions peculiar to crushers and equipment into three categories :

- Primary and secondary, jaw and gyratory crushers (gouging abrasion)
- Ball and rod mills (grinding abrasion)
- Abrasive wear caused by mineral particles transported in a fluid (erosion)

A fourth category can be added : the impact crushers grinding by impact at high speed wich have very different rates of wear.

The technical literature abounds with the results of abrasion tests carried out in situ or in the laboratory, with a number of these not being very trustworthy. Measurements taken in situ are very scattered, and few correlations have been established between laboratory tests and service wear.

1. Jaw and Gyratory Crushers

Utilization time is the criterion most often used to characterize the abrasion resistance of crusher liners, the utilization times being distributed according to a Gaussian law with a large dispersion, (the variation coefficients (standard deviation/mean) being of the order of 16% for secondary crushers used in quartzite crushing (figure 140). The dispersion of utilization times is due to the heterogeneity of the mineral

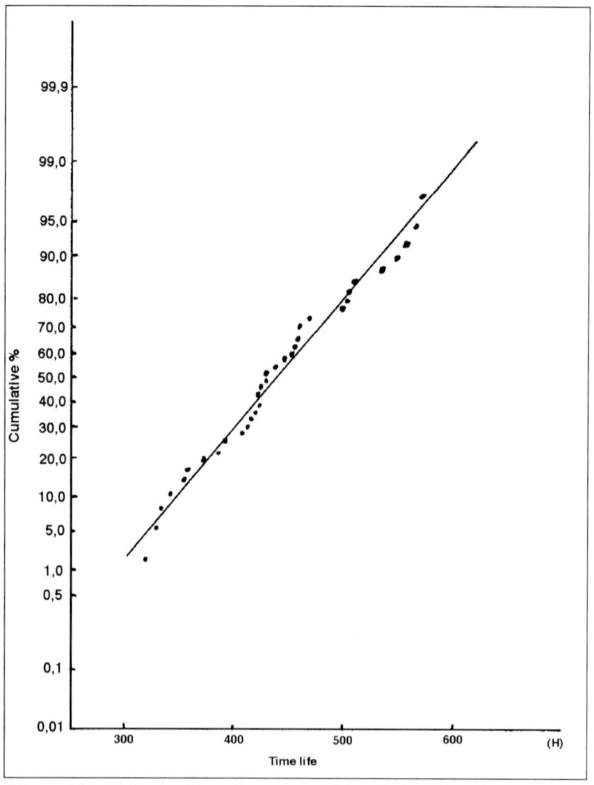

Figure 140 : Galtonian Transformation of Cumulative Utilization Time for Secondary Gyratory Crusher Liners (48') used for Crushing Quartzite (177)

deposit and to parameters related to the liners and conditions of use. As an example of the effect of heterogeneity of the deposit, consider the 18% increase in global consumption of wear components, which occurred in 1976 at the Colorado Climax Mine (USA), following the working of a zone of higher mineralogical hardness.

Although for a given component suppliers produce pieces of the same geometry, it should be noted that for the same supplier, variations in weight for the combined mantle and bowl liner are of the order of ± 1.7%. Variations are greater from one producer to the other, being 10.2% for the mantles, and 7.6% for the bowl liners.

The presence of micro-shrinkage zones localized at the interface between two solidification zones is frequently visible on partially worn liners. They have, beyond any doubt, a negative effect on abrasion resistance.
The specification ranges of chemical composition are relatively wide, so as to benefit from low manufacturing costs ; on the other hand, the same ranges can cause variations in abrasion resistance.

A service fracture can shorten the utilization time of the liners. Some fractures are associated with the presence of micro-shrinkage, porosity, sand inclusions or hot cracking. Others, which propagate in zigzags, are at times produced in otherwise sound castings, with good toughness and a chemical composition conforming to specifications.

The fracture can be caused by the passage of an uncrushable body through the mill, and occasionally when the resin filling of the interstices between the liner and its support is inadequate. In this case a circumferential fracture can be seen near the machined part supporting the liner. The fracture of the bowl liner fixation hooks can also be the cause of a breakdown.

Insofar as these fractures are not produced after a very brief period of use, shortened utilization times risk being counted as a normal life span and can lead to the assumption of a lower abrasion resistance.

The lifetime of liners can be prolonged or shortened to a considerable extent according to conditions of utilization, but the latter are difficult to characterize and even more difficult to quantify.

Conveyor belts are not equipped with an automatic weighing device, and feeding rates are simply evaluated. The distribution between mills working in parallel is apparently identical in volume and particle size distribution but the flow can be different.

The setting conditions for the opening of the mill may have to be modified during utilization to obtain the desired product. The individual nature of each operator introduces a random variable, and the intervention of different operators is an inevitable cause of dispersion.

The liners are changed following a decision of the maintenance personnel, who seek to ensure maximum production and to minimize intervention periods. The criteria which govern the operator's decision to reduce the risks of deterioration of the equipment are as follows :

- The presence of a small crack capable of propagating and causing fracture of the liner ;
- Wear of the liner ascertained after visual examination ;
- A wear rate too advanced to allow use, during the weekend, or to wait for the following maintenance operation ;
- When the liner is unable to ensure efficient crushing.

The premature exchange after 24 hrs. of a liner, in the example shown in figure 140, represents a difference of 5.4% in relation to the average time of use.

Liners can be damaged for mechanical reasons peculiar to the equipment and independent of the liners e.g. shaft fracture. The mantle and bowl liner are used in pairs, but when a mechanical event leads to only one of these being damaged, the other can be used if its wear rate is low. In this case, at the next dismantling operation, one of the two pieces will be incompletely worn, whence a wear rate is calculated by excess.

It can be calculated from the data base of figure 140 that a difference of average time of use between two grades of less than 25%, (calculated after testing four liner combinations by grade, that is, eight liner combinations in total), is not significant at a probability level of 95%.

The weight loss per crushed ton should be a better criterion than utilization time, for characterizing the abrasion resistance of crusher liners.

A study of the effect of metallurgical factors on the abrasion resistance of manganese austenitic steels, in the crushers under consideration, is better carried out using a small jaw crusher and a well-known mineral, available in sufficient quantities (178) (179) (73).
This method was used to study the effect of chemical composition of manganese austenites on abrasion resistance using a quartzite; the minerals treated being represented in large part by siliceous rocks (73). The work-hardened layer of trial crusher liners would be especially differentiated from that of commercial crusher liners of larger sizes, by a lower thickness (figure 141). As surface hardness is a significant parameter according to Torrance's model, it can be considered that the results obtained following this method give a good idea of the effect of metallurgical factors on abrasion resistance in the machinery and with the mineral under consideration.

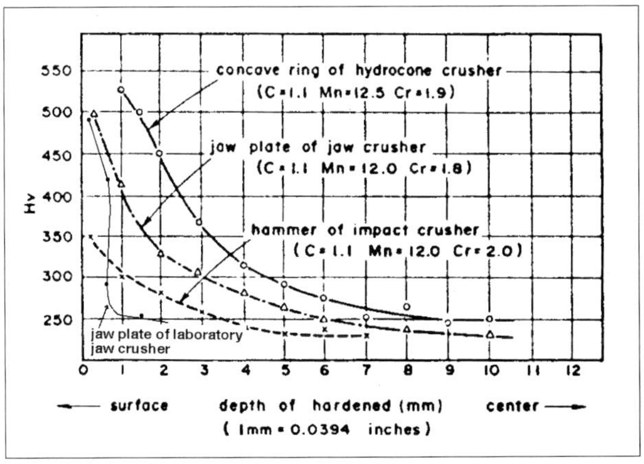

Figure 141 : Work-hardened Hardness of Austenitic Manganese Steel Castings. Commercial Crushers (19).
Test Crushers (73)

Storage and consumption of the mineral being cumbersome for systematic testing, a pin-test on less costly abrasive paper is usually used. A correlation has been established between the two methods whereby the logarithm of the relative wear rate, measured after crushing quartzite, is proportional to the wear rate measured in the pin test on 180 mesh alumina abrasive paper (figure 142).

The abrasion resistance of manganese austenites, measured using both a jaw crusher and a pin test, increases monotonically when carbon content increases and manganese content decreases (figures 143-144).

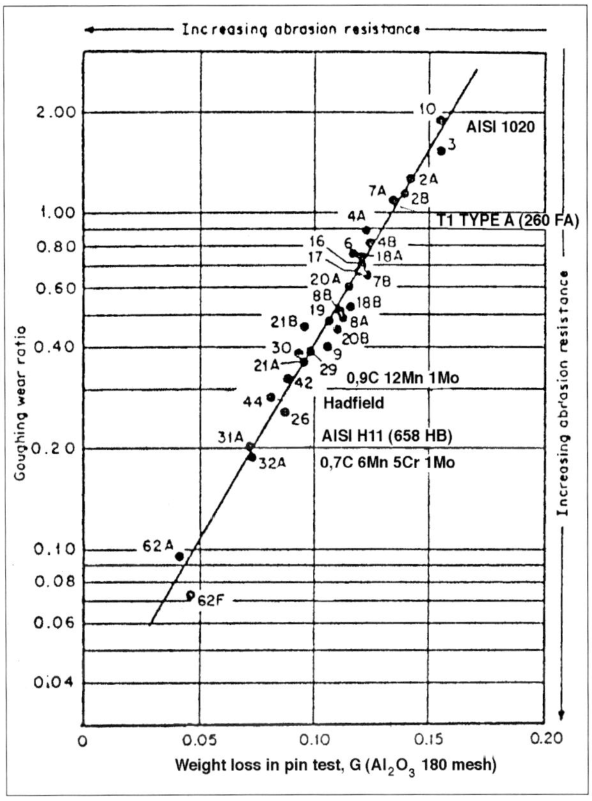

Figure 142 : Relation between Relative Wear Measured after Crushing 960 kg of 40-50 mm Quartzite an Pin Test Wear (180)

Pin test abrasion, in unspecified conditions, of manganese austenites containing 0.5 to 1.0% carbon and 12.0 to 18.0% manganese (figure 145), show that curves of equal abrasive wear rates have a 'V-form', as the equal elongation curves in the Mn-C plane, below a manganese content limit. At a manganese content above this limit the same wear rate can be observed at two different carbon contents. A similar wear rate is obtained with 14% manganese and 0.78 or 0.98% carbon. The relative abrasive wear rate at the cusp of the V increases with an increase from 0.285 to 0.325 when the manganese and carbon

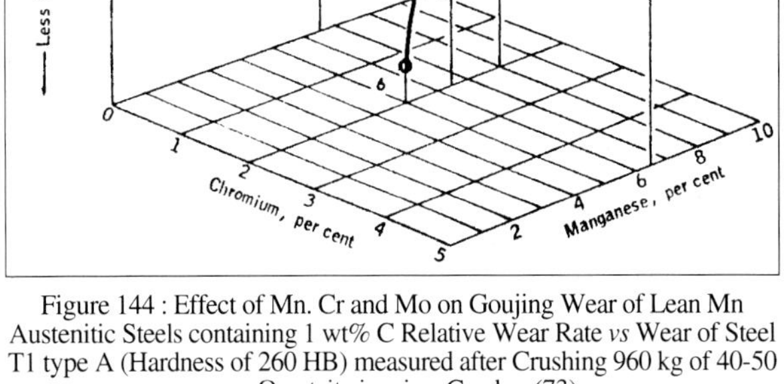

Figure 144 : Effect of Mn. Cr and Mo on Goujing Wear of Lean Mn Austenitic Steels containing 1 wt% C Relative Wear Rate *vs* Wear of Steel T1 type A (Hardness of 260 HB) measured after Crushing 960 kg of 40-50 mm Quartzite in a jaw Crusher (73)

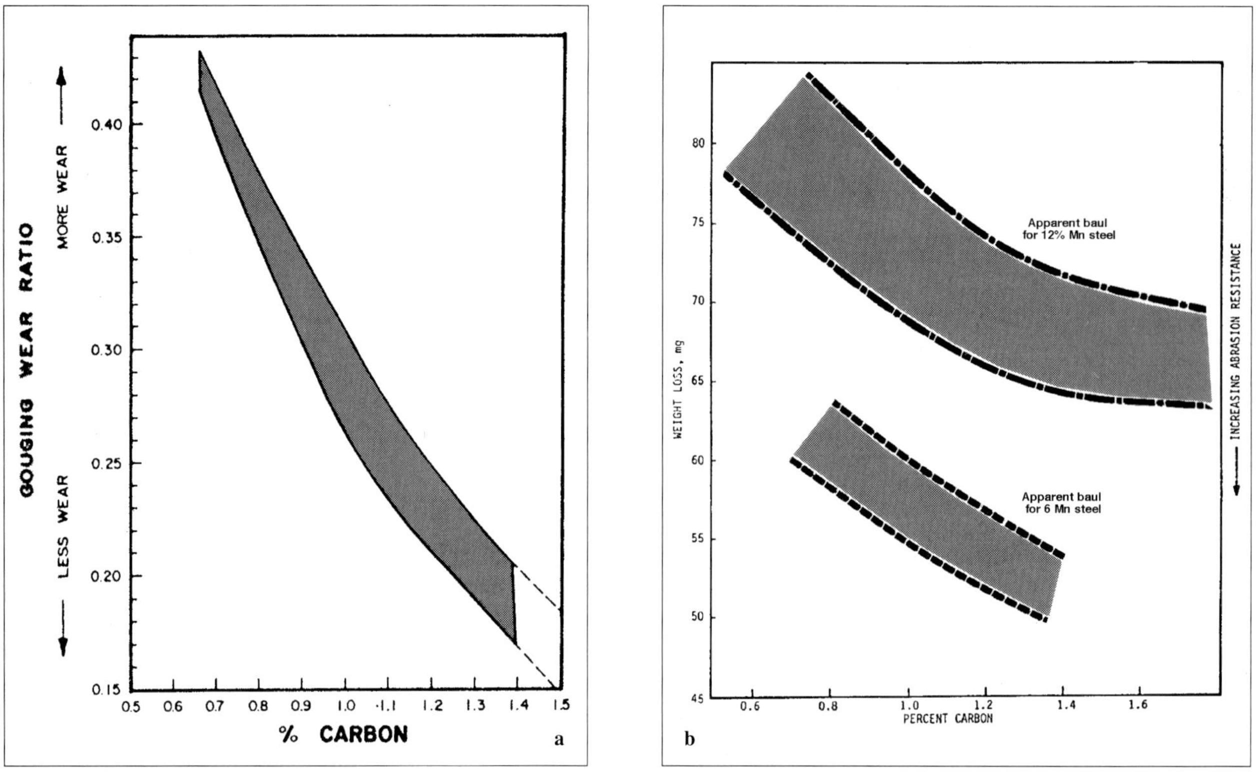

Figure 143 : Goujing Wear Ratios of Austenitic 12% Mn Steels vs C Content. - a. Test jaw crushers (179) b. Pin test (181)

contents change from 12.4% Mn and 0.81% C to 17.5% Mn and 0.92% C respectively.

The shape of the curves of equal abrasion resistance is a result of the microstructure of the work-hardened state. Certain austenites are metastable and can be distinguished from each other by the microstructure of thc induced martensite, whilst others are stable. These curves show that metastable austenites with high manganese content and low carbon content can have good abrasion resistance. Table 22 brings together a few test results, with a view to evaluating the effect of grain size and of some alloying elements on the abrasion resistance of Hadfield steel. Taking into account the effect of carbon content on abrasion resistance, it can be noted that, (with the exception of the addition of 3.5% nickel, which lowers abrasion resistance from 12 to 18%), an addition of 2% chromium and of 1 to 2% molybdenum leaves the wear rate value within the dispersion band of the wear rate-carbon percentage relation, of the steel with 12% manganese. Table 22 also shows that an addition of 0.1 to 0.125% of zirconium has no significant effect on the abrasion resistance of the liners of gyratory crushers used for crushing auriferous quartzite, and that a difference in grain size, of two orders of magnitude, has a negligible effect on abrasion resistance.

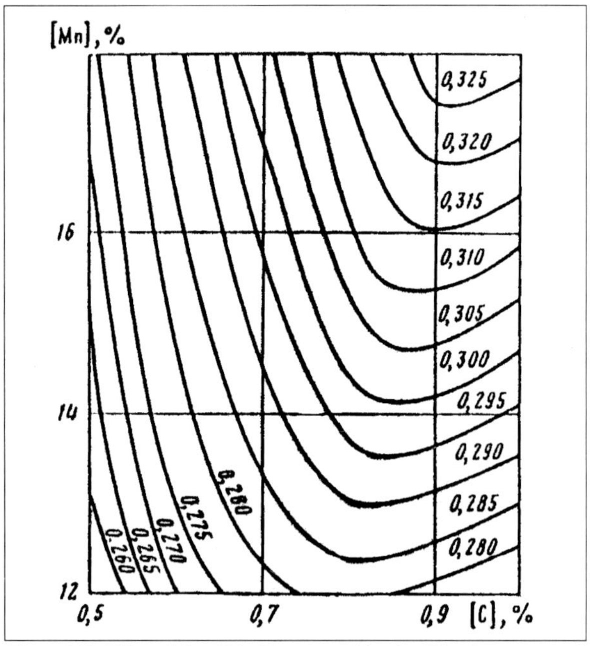

Figure 145 : Effect of C and Mn Contents on Abrasive Wear Rate measured by Pin Test (182) (type of abrasive unspecified)

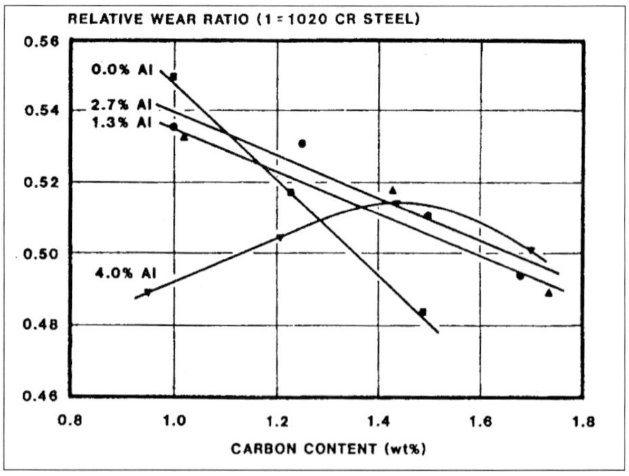

Figure 146 : Effect of C and Al on Pin-abrasion Test Relative Wear Ratio

Chemical Composition %							Grain Size (mm)	Wear Rate*
C	Mn	Si	Cr	Mo	Ni	V		
1.0	11.91	0.37			0.36			0.279 0.280
0.99	12.65	0.64			3.54			0.328 0.292 +12%
1.18	12.64	0.59	2.14					0.249 0.231 +8%
1.28	12.8	0.60		0.96				0.200 0.212 -5.6%
1.07	12.71	0.59		2.05			2.2	0.253 0.248 +2%
1.07	12.71	0.59		2.05			0.024	0.246 0.248
0.30	13.8	0.60		1.80	3.10	0.61		0.629
0.60	13.8	0.60		1.80	3.10	0.61		0.535
0.90	13.8	0.60		1.80	3.10	0.61		0.381 +18%

Table 22 : Effect of the Addition of Allowing Elements, and of Grain Size, on Abrasion Resistance of Hadfield Steel and Low Carbon Austenites (73)
* Relative Wear rate involving T1 Type A steel, measured after crushing 960 kg of 40-50 mm quartzite

Pin test trials carried out under experimental conditions identical to those used by D.E. Diesburg, F. Borik and H.L. Arnson (180) (181), show that an addition of 4% aluminium improves abrasion resistance at carbon contents below 1.25% (figure 146).

The dispersed carbides grade, obtained by heat treatment at 600°C followed by a partial redissolution in the temperature range 980 to 1010°C (ASTM norm A128 - grade E2) can give a better performance than standard Hadfield steel (figure 147), but its behaviour in service appears erratic, and no improvement has been observed at specific sites (table 23). The effect of the cycle of heat treatment and spheroidizing of carbide platelets, applied to Hadfield steel with 1.10% carbon and 13% manganese, on abrasion resistance measured with the pin test (test conditions are unspecified) is illustrated by figure 148.

Chromium, even at contents of up 5%, has no significant effect on the abrasion resistance of metastable austenites, in contrast to molybdenum, which brings about an improvement (figure 144).

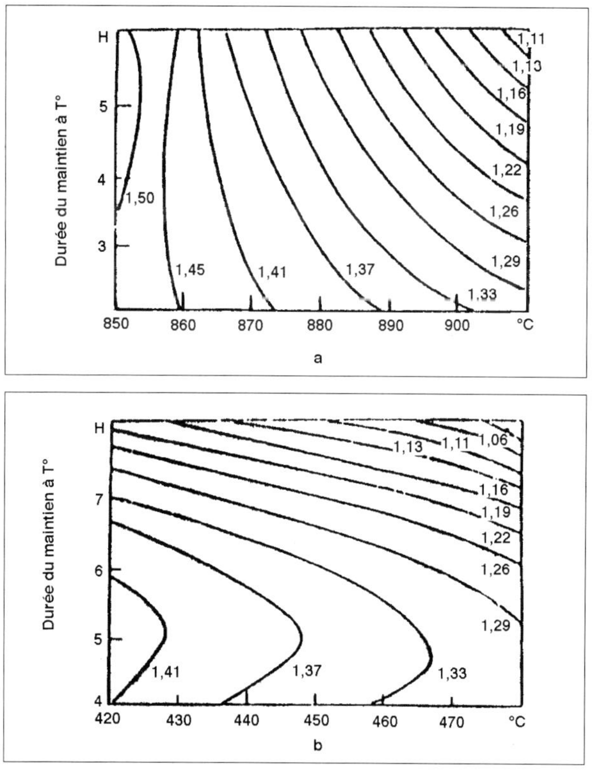

Figure 148 : Effect of Carbide Precipitation and Spheroidization on Abrasion Resistance of Hadfield Steel (1.10 C - 13 Mn) (184)
a.Effect of Spheroidization Temperature and Maintaining Time on Abrasion Resistance (Carbide Precipitation 450°C - 6hr)
b.Effect of Carbide Precipitation Temperature and Maintaining Time on Abrasion Resistance (Spheroidization Treatment 880°C - 4 hr)

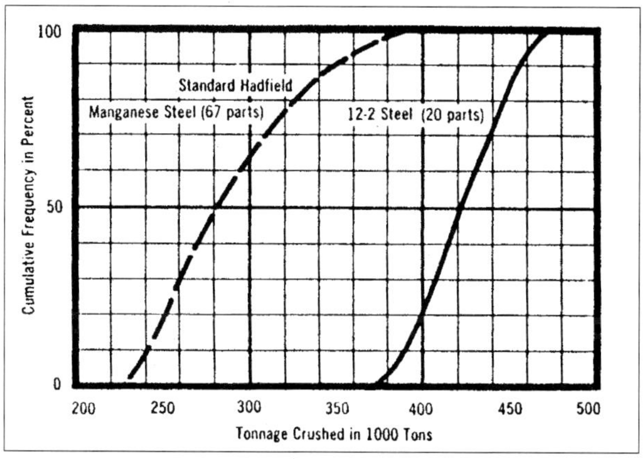

Figure 147 : Frequency Curves Showing the Maximum Tonnage Crushed with Bowl and Mantle Liners of Symons 2.31 m (7 ft) Secondary Cone Crushers made of Standard Hadfield Mn Steel and of 12 Mn-2 Mo 1.10 to 1.20% C Steel containing Dispersed Carbides (WQ + 593°C + 980°C) (75)

	Average Plant (kgT^{-1}) x	Standard Deviation (kgT^{-1}) σ	Number of Tests n
Hadfield	0.02005	0.0050	10
0.1 - 0.125% Zr	0.01967	0.0040	4
1.25%Mo Carbide Dispersion Hv = 241 - 269	0.0199	0.0044	4
Total	0.01993	0.0044	18

Table 23 : Effect of the Addition of Zirconium, and of Carbides Dispersion, on the Abrasion Resistance of 4 1/4 Gyratoy Crusher Liners in a South African Gold Mine (Western Deep Level Gold Mining Co. Ltd.) (183)
$\sigma/x = 22\%$

2. Impact Crushers

At high speed the hammers and impeller bars of impact crushers undergo deformation of the surface layers, which is the reverse of jaw and gyratory crusher liners, where the deformation can be termed static. The surface work-hardening of hammers and impeller bars seems less than that of liners for other types of crushers (figure 141). The plastic deformation during the formation of microchips is localized in the shear zone. The temperature in this zone increases and can initiate a recovery which neutralizes the work-hardening. In extreme deformation conditions, shearing occurs adiabatically, and the quantity of heat generated produces a catastrophic softening effect.

The behaviour of abrasion-resistant materials in impact crushers is more complex than in jaw and gyratory crushers. The classification of materials by their abrasion varies according to the rate of impact.

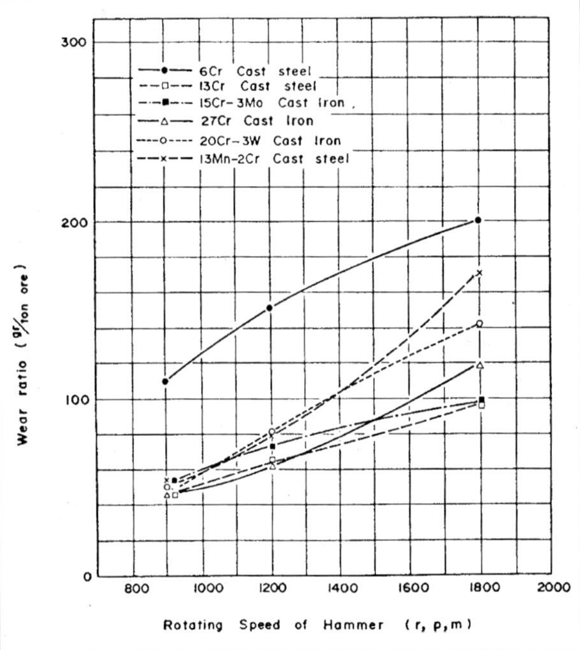

Figure 149 : Influence of Rotating Speed on Wear Rate of Hammers of an Impactor Crusher, after Crushing a ton of 80-20 mm granite containing 60% SiO_2 (19)

Figure 149 shows the effect of the rotation speed of a hammer mill crusher on the wear rates of different abrasion-resistant materials used for crushing granite. All the abrasion-resistant materials studied, with the exception of steel at 6% chromium (probably low carbon) have practically the same abrasion resistance at low rates of impact. Wear rate increases when the impact rate increases and the materials can be differentiated : cast iron with 15 Cr - 3 Mo and steel with 13% chromium have the lowest wear rates ; steel with 6% chromium show the worst abrasion resistance ; cast iron at 20 Cr - 3 W and Hadfield steel are ranked in an intermediary position between 6% Cr steel and 13% Cr cast steel.

The study of the effect of metallurgical factors on the abrasion resistance of manganese austenites in impact crushers, has been the subject of only a few publications, and experimental conditions are insufficiently precise to define the conditions of commercial applications. The test results analysed hereafter are thus only representative of the experimental conditions under which they were obtained.

The abrasion properties of the hammers of an impact crusher of manganese austenite used for crushing asbestolite, increases with an increase in carbon content and when the manganese content decreases (figure 150). The effect of adding 2% chromium on

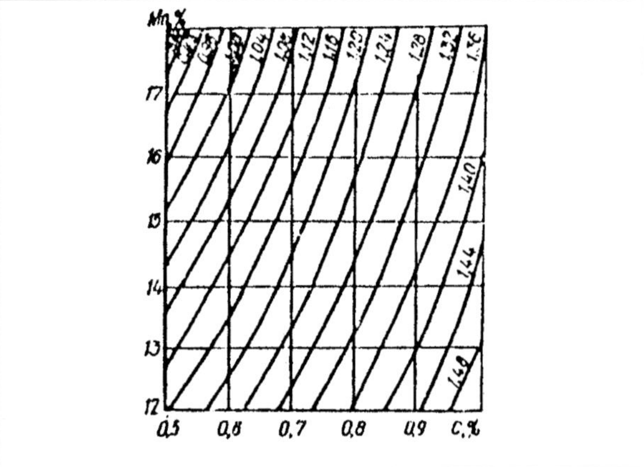

Figure 150 : Effect of Mn and C Contents on the Tonnage of Asbestolite Necessary to wear 1 mm of the Hammers of an Impactor Crusher (Abrasion resistance in Relative values) (184)

the wear resistance of the Hadfield steel Hazemag crushers' (AP 4-80) impeller bars was found to be insignificant in crushing coal, but brought about a significant improvement in the crushing of a mixture of coal and schist (table 24). An addition of 2% chromium and 0.15% titanium to a Hadfield steel was found to have a beneficial effect on the abrasion resistance of impeller bars of an impact primary crusher in the crushing of limestone and granite (186).

		Average Lifetime (H)	Standard Deviation	N° of Measuremts	t	$t_{0.95}$	
Coal Crushing 40-50 at 1-10 (mm)	Hadfield	1343	414	7	1.55	1.79	NS
	2% Cr	1648	291	6			
Coal & Schist Crushing	Hadfield	1052	396	9	5.21		VS
	2% Cr	1977	313	7			

Table 24 : Effect of the Addition of 2% Cr on Abrasion Resistance of Hazemag AP 4-80 Impeller Bars (185)

3. Ball and Rod Mill Crushers

The different elements of the liner of a ball or rod mill crusher do not wear at the same rate. Figure 151 shows as an example that the wear of the lifting bars of a rod mill crusher is as rapid as the other shell liner components.

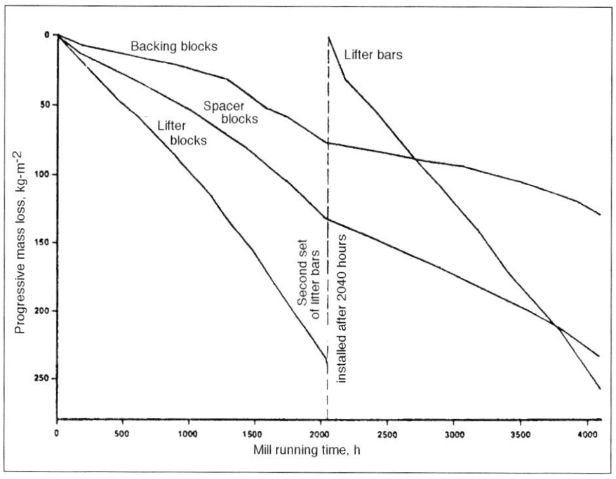

Figure 151 : Progressive Wear Rates of Manganese Steel Liners
(F = 2.348 m ; L = 3.68 m - Crushing of auriferous quartzite) (183)

The abrasive wear mode does not seem to be different, and T. Normans (187) has found in a study of abrasive wear with 150 mm diameter balls, that the results obtained were representative of the mode of wear of this equipment.

Metallurgical factors have qualitatively the same effect on the abrasion resistance of manganese austenites in the equipment under consideration, as for jaw and gyratory crushers. Abrasion resistance increases when carbon content increases, and when manganese content decreases (figure 152).

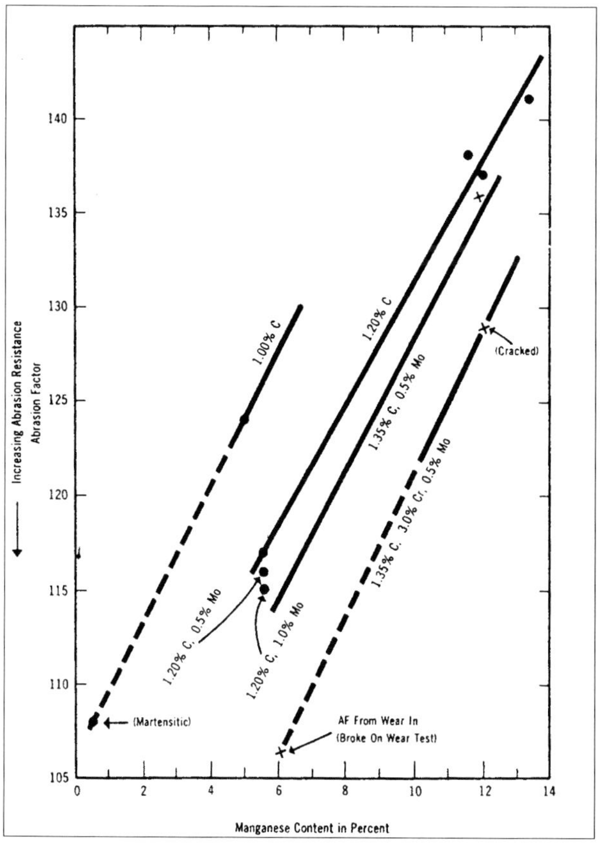

Figure 152 : Influence of Mn Content on the Abrasion Factor of Austenitic Steels with Various C and Alloy Contents (188)
(Crushing of Quartzite with 150 mm-diameter balls)

Copper, which stabilizes the austenite, has an unfavourable effect on the abrasion resistance of metastable austenites. Silicon has a similar effect. This is in contrast with molybdenum, which destabilizes the austenite and has a beneficial effect on the abrasion resistance of these austenites (figure 153).

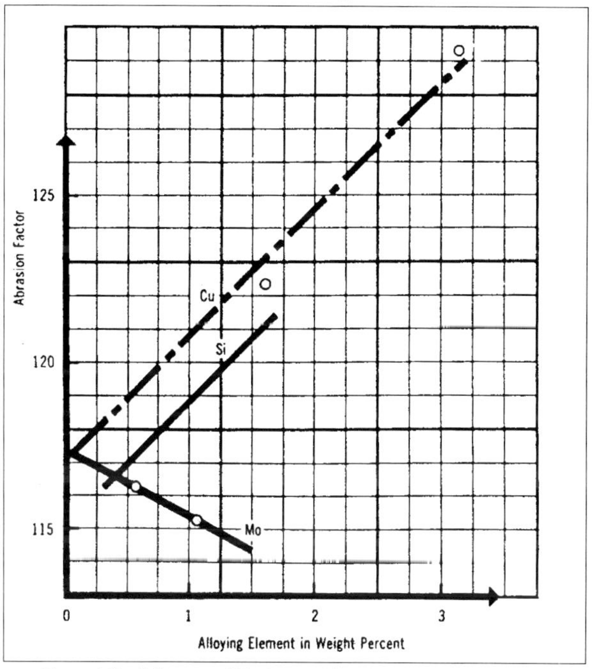

Figure 153 : Influence of Cu, Mo and Si Additions on the Abrasion Resistance of an Austenitic 6% Mn 1.2% C Steel (75)
(Crushing of Quartzite with 150 mm-diameter balls)

d. Erosion

Erosion is caused by the impact of hard and abrasive particles. Speed, mass, angle of impact of the abrasive, its sharpness and its friability, are all very important factors. Conditions are even more complex in commercial machinery, where particle size, kinetic energy, and the angle of impact of the particles, are

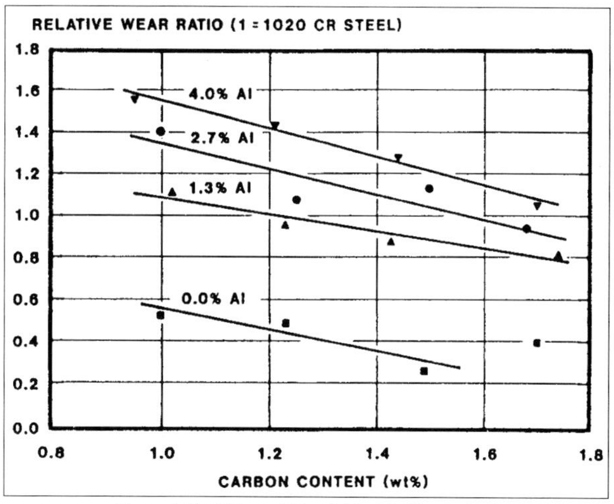

Figure 154 : Effect of C and Al on Rubber-wheel Abrasion Test Relative Wear Ratio (148)

Tension SAE 1020
Abrasive Silica 50-70 mesh
Load 22.7 kg
Rubber-wheel Hardness 60 shore A

Impingement Angle	Mild Steel	Steel 12% Mn - 1.2 C with addition of :							
		Hadfield base	Nb	V	Mo	Cr	W	3Cr	3W
30°	222.2	100.1	118.0	112.8	106.8	112.6	74.9	103.4	65.0
45°	179.5	40.8	30.3	40.3	29.3	33.6	35.0	25.8	35.4
60°	165.0	62.9	70.7	37.4	60.8	64.5	35.6	59.8	38.0
90°	67.9	30.2	20.3	25.0	20.9	23.8	10.8	17.0	11.4
Σ Performance (%)	37	100	98	109	107	100	150	114	156

Table 25a : Erosion Tests with Clinker at 75 m/sec (loss in mg/kg abrasive) (particle velocity = 75 m/sec ; particle size : 0.7/1 mm ; abrasive : clinker ; wear : mg/kg abrasive)

Impingement Angle	Steel 12% Mn - C with addition of :						
	Hadfield	1Cr 1.2C	3Cr 1.4C	6Cr 1.8C	1W 1.2C	3W 1.4C	6W 1.8C
30°	89.3	73.0	69.7	72.6	75.7	68.4	69.8
45°	104.1	105.7	102.4	112.3	107.9	95.7	93.1
60°	109.5	110.3	110.1	117.0	111.8	110.8	115.1
90°	88.6	92.6	89.3	95.0	94.6	97.0	104.9
Σ Performance (%)	100	103	105	99	100	105	102

Table 25b : Erosion Tests with Quartzite at 50 m/sec (loss in mg/kg) (V = 50 m/sec ; abrasive : SiO_2)

Condition	Hardness, HRC	Pin Test Weight Loss, g		Rubber Wheel Abrasion Test Weight Loss, g[a]
		Alumina 180 mesh (80 μm)	Garnet 150 mesh (100 μm)	
Austenitic	18.0			0.66
Austenitic	17.3			0.66
Hardened	61.3			0.30
Hardened	61.1			0.30
Austenitic	15.3	0.0538	0.0460	
Austenitic	15.8	0.0551	0.0467	
Hardened	61.6	0.0767	0.0682	
Hardened	60.9	0.0781	0.0688	

Table 26 : Results of Hardness Determinations and of Pin and Rubber Wheel Abrasion Tests made on C alloy Steel* (189)

[a] Weight loss during 5000 Revolutions at Shore Durometer Hardness of 55

* C% - 1.15, Mn% - 5/5, Cr% - 1.5, Mo% - Si% - + --------------------------1.1, Co% - 1.5

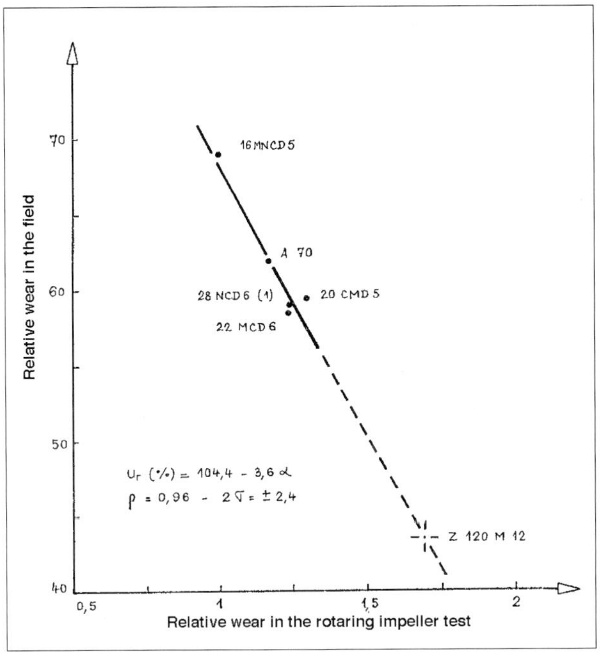

Figure 155 : Relationship between the Abrasion Resistance of Abrasion-resistant Steels and a Hadfield Steel, using fresh concrete at 300 kg measured during use and in the rotating impeller test (190)

not homogenous. The effect of chemical composition on the erosion resistance of manganese austenites has been the subject of several studies using different erosion test methods :

1)by pure abrasion on a machine with a rubber wheel ;
2)by controlled impact of abrasive particles ;
3)by an abrasive in a machine where the test piece ensures the continuous stirring up of the abrasive.

Erosive wear of austenites with 12% manganese, caused by silica in a machine with a rubber wheel, decreases according to a linear relation when the carbon content increases. An addition of 1.3% aluminium doubles the wear rate (figure 154). Additions of Nb, V, Mo, Cr, and W have little effect on erosion resistance to silica, but on the other hand an addition of 3% chromium, and especially 3% tungsten has a beneficial effect on clinker erosion resistance (table 25).

In certain commercial equipment subjected to erosive wear, surface work-hardening during service is insufficient to benefit from the advantages of Hadfield steel. In these installations, metastable austenitic steels, treated to obtain a martensite-austenitic structure with carbide, (which enables their erosion resistance to be doubled in relation to the austenitic structure), can be a solution. It is interesting to note (table 26) that the effect of the microstructure on erosion resistance is reversed in relation to the pin test abrasion resistance, where the abrasion resistance of the austenitic structure is higher by 50% than that of the martensite-austenitic structure.

Service behaviour can at times be evaluated with a simple apparatus like the one in which the test piece ensures the stirring up of the abrasive. Figure 155 shows that the erosion resistance of Hadfield steel during cement mixing is greater than that of a steel with 0.7% carbon and low alloy steels selected for their abrasion resistance.

V. Armour Plates

Hadfield steel is used in the manufacturing of helmets and security armour lining, in order to ensure protection from projectiles with a low impact rate. For this application it has been found that work-hardening by deep drawing (for helmets) or by cold rolling, were beneficial. Figure (156) gives, according to Woodward (191), an outline diagram of the three main means of perforation of a homogenous lining, after impact of a projectile.

The critical thickness of a perforation-resistant plate according to the ductile hole formation (figure 156a) obeys the following equation :

$$h = (mV^2/\pi D^2)\ (^1/_\sigma) \qquad (1)$$

where h = target thickness
V = projectile speed
D = projectile diameter
σ = resistance coefficient of the target material
m = mass

The coefficient σ is determined experimentally under compression, at a high deformation rate ($\varepsilon = 1.0$) given the extent of plastic deformation during perforation.

This equation is valid over a wide range of characteristics of the target (strength and thickness) and of the projectile (diameter, mass and speed). It can also be satisfactorily applied to two other modes of perforation.

The calculation implies that the projectile is undeformable and its trajectory to the target surface is perpendicular. For oblique trajectories, the apparent thickness beneath the angle of trajectory considered can be taken as the value of critical thickness, to a satisfactory approximation. When the projectile is deformed during impact, as with lead bullets with or without brass casings, the calculated thickness is considerably over-estimated.
The choice of an armour plate does not rely solely on the value of the critical perforation thickness h, but

Material	Hardness (HV5)	Strength σ_0(MPa)	Density ρ(Kg/dm^{-3})	Cost κ* ($/Kg)	Thickness Factor $1/\sigma_0$(MPa^{-1}) x 10^{-3}	Weight Factor ρ/σ_0 x 10^{-3}	Cost Factor $\kappa\rho/\sigma_0$ x 10^{-3}
Hot rolled CS 1020 mild steel	136	976	7.83	0.24	1.025	8.02	1.92
SAE 4130 Steel (Quenched and Tempered)	300	1240	7.83	2.01-2.65	0.806	6.31	12.7-16.7
	350	1420	7.83	2.01-2.65	0.704	5.51	11.1-14.6
	400	1600	7.83	2.01-2.65	0.625	4.89	9.8-13.0
P8 Hadfield steel (hot rolled)	194	1690	7.87	1.02	0.592	4.66	4.75
5083 Aluminium Alloy(cold rolled)	105	452	2.80	1.90	2.21	6.19	11.8
7039 Aluminium Alloy (Aged)	155	638	2.80	--	1.57	4.38	--
Titanium-Com mercial Purity (Mill-annealed)	160	1167	4.54	12.2	0.857	3.89	47.5
Titanium 318 Alloy (Forged and annealed)	315	1685	4.42	33.05	0.593	2.62	86.6
Titanium 8 Al-1 Mo-1 V alloy(hot Rolled-Duplex annealed)	315	2020	4.42	--	0.495	2.19	--

Table 27 : Comparison of Homogenous Armour Materials. *Data obtained June 1976

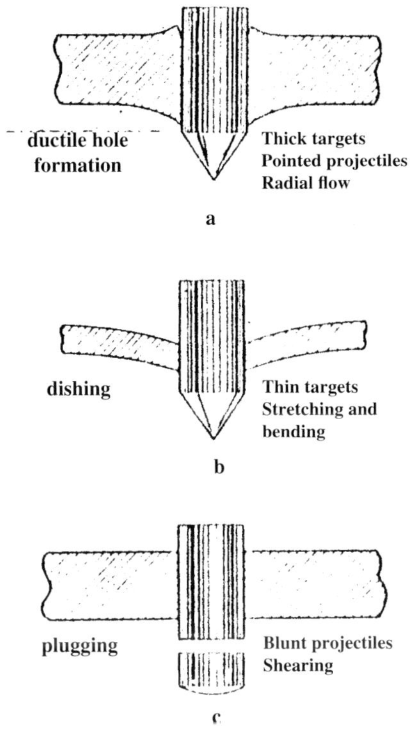

Figure 156 : Schematic Illustration of the 3 Principal Failure Modes Associated with the Penetration of Homogenous Metal Armour by Armour Piercing Projectiles

also on the weight P = ρh, and the cost C = Kρh (per unit mass of target area is κ times Mo moss per unit area).

The calculated values are generally 20% to 35% higher than the experimental ones, the differences arising from the fact that the formula used does not take into consideration secondary effects, such as rubbing or inertia. In order to proceed with a relative classification, it is sufficient to minimize the values 1/σ for thickness, r/s for unitary weight, and Kρ/σ for the price.

Table 27 compares different alloys for armour plate : it can be seen that an excellent ranking is given to Hadfield steel. The price factor of Hadfield steel is the lowest of all, and only the alloy Ti - 8 Al - 1 Mo - 1 V has a lower thickness factor. Titanium alloys naturally have a lower weight factor, but their utilization in thin section is limited, as with the steel SAE 4130 to give a hardness greater than 350-400 HB, because of a strong decrease in perforation resistance, due to the appearance of adiabatic shearing as a means of deformation.

References

1. Thermodata, 13.01.1988, Ref. PYC / mm / 88.7, pp. 1-11 (Report for IMI)

2. E. Schürmann, I.K. Geissler, Giesserei-forschung, 1977, 29 (4) pp. 153-159

3. V.G. Rivlin, Inter Metals Review, 1984, Vol. 29 (4) pp. 299-327

4. A. Hellawell, Equilibrium Diagrams, 1956 (26) The Institute of Metals, London

5. R. Benz, J.F. Elliot, J. Chipman, Metall. Trans., 1973 (4) pp. 1449-1452

6. S. Schenk, M.G. Frohberg, E. Steinmetz, Anch. Eisenhüttenwes., 1963, 34, pp. 37-42

7. B.J. Nikolin, L.I. Lysak, Yu. N. Nakogon, Fizika Metallov. Metalloved., 1971, 32, pp. 871-874

8. H. Schumann, Neue Hutte, 1972, Oct., 17 (10) pp. 605-609

9. T. Kato, M. Fujikura, K. Ishido, A. Kimura, Y. Takeuchi, N. Kawasaki, S. Yahagi, Trans. I.S.I.J., 1981, Vol. 21, pp. 852-862

10. I. Schmidt, (Proc. Conf.) Int. Conf. Martensitic Transformations, Japan Inst. of Metals, 1986, pp. 497-502

11. N. Jost, I. Schmidt, (Proc. Int. Conf.) Wear Materials, ASME, 1985, pp. 205-211

12. Ö.A. Atasoy, Z. Metallkde., 1984, Bd. 75, H6, pp. 463-471

13. J.M. Zimmer, V.D. Bailey, U.S. Patent, 12 Sept. 1989, Patent N° 4 865 662

14. K. Kinoshita, Y. Yoshii, H. Kitahoka, A. Kawaharada, K. Kishikawa, O. Tanigawa, J of Metals, 1984, March, pp. 38-43

15. J. Tasker, Intermountain Minerals Symposium, Vail, Colorado, 1982, 3-6 Aug., pp. 3-19, Ed. Climax Molybdenum Co.

16. F. Fremunt, K. Stransky, J. Velicka, Archiv. Eisenh., 1972, Vol. 43 (10) pp. 769-774

17. M. Slesar, L. Kamaras, Slevarenstk. 1968 (1) pp. 18-21

18. G. Collette, C. Crussard, A. Kohn, J. Plateau, G. Pomey, M. Weisz, Rev. Metall., 1957, June, pp. 433-486

19. N. Tsujimoto, A.F.S. Int. Cast Metals J, 1979, June, pp. 62-77

20. F. Maratray, A.F.S. Trans. Modern Casting, 1964, Vol. 72, pp. 1-12

21. T. Saito, Iron and Steels, 1972, Vol. 58 (3) pp. 123-133

22. K. Sato, T. Tagawa, Y. Inoue, Materials Science Engineering, 1989, A111, pp. 45-50

23. R.K. You, Po-We Kao, D. Gan, Materials Science Engineering, 1989, A117, pp. 141-148

24. J. Le Coze, Cercle d'Etude des Metaux, 1977, Comm. N° XI, pp. 1-17

25. P.H. Adler, G.B. Olson, W.S. Owen, Met. Trans. A, 1986, Oct., Vol. 17A, pp. 1725-1737

26. K.S. Raghovan, A.S. Sastri, M.J. Marcinkowski, Trans. A.S.M., 1969, July, Vol. 245, pp. 1569-1575

27. H.J. Lai, C.M. Wan, J Materials Science, 1989, 24, pp. 2449-2453

28. V.A. Teplov, L.G. Korshunov, V.A. Shabashov, R.I. Kuzneshov, V.P. Pilyuzin, D.I. Tunisha, Met. Metalloved., 1988, Vol. 66 (3) pp. 563-571

29. D. Rittel, I. Roman, Materials Science Engineering, 1989, A110, pp. 77-87

30. C.H. White, R.W.K. Honeycombe, J Iron and Steel Institute, 1962, June, pp. 457-466

31. Y.N. Datur, W.C. Leslie, Metal Trans., 1981, May, Vol. 12A (5) pp. 749-759

32. S.A. Sastri, R. Ray, Metal Trans., 1974, June, Vol. 5, pp. 1501-1503

33. V.A. Shabashov, L.G. Korshunov, Yu. V. Baldokin, Fizika Metallov. I. Metalloved., 1989, Vol. 67 (6) pp. 1197-1203

34. K.I. Krasikov, L.P. Strok, N.K. Shaurova, Metalloved. Tern. Obrab. Metallov., 1977 (3) pp. 7-10

35. H. Schumann, Neue Hütte, 1972, Oct. (10) pp. 605-609

36. F. Gauzzi, M. Rossi, B. Verdini, Metallurgia Italiana, 1971 (11) pp. 555-558

37. M.A. Filimov, V.A. Konilova, V.E. Luzovik, Cernaya Metallurgia, 1988 (6) pp. 75-79

38. C.H. Shih, B.L. Averbach, M. Cohen, Landolt Börnstein, 1963, Vol. A Part 29, 863, Springer Verlag

39. P. Guillaume, A. Magnée, D. Coutsouradis, L. Habraken, Mem. Scient. Rev. Metal., 1978, March, 137

40. A. Magnée, Bull. Cercle d'Etudes des Metaux, 1984, March, Vol. 15 (5) Conf. N° 15, pp. 1-12

41. G. Collette, C. Grussard, A. Kohn, J. Plateau, G. Pomey, M. Weisz, Rev. Metall., 1957, Vol. 54, pp. 433-481

42. J.H. Hall, Trans. A.I.M.E., 1929, 84, pp. 382-427

43. S.V. Mamayev, L.N. Romashev, D.P. Rodionov, Fizika Metal. Metalloved., 1989, 67, 909 (trans. Eng. : Phys. Met. Metall., 1990, Vol. 69 (5) 46-51)

44. S.V. Mamayev, L.N. Romashev, D.P. Rodionov, D.I. Tupika, G.G. Talents, Phys. Met. Metall. 1990, Vol. 69 (5) pp. 46-51

45. N.Ya. Samarin, R.Z. Kats, V.A. Startsev, I.A. Petrova, Metallov. Obrab. Metallov. 1981, (1) pp. 39-42

46. V.N. Krivobok, Trans. Amer. Soc. Steel, 1929 (15) pp. 893-928

47. P. Chévenard, Met. Corr. Ind., 1935 (10) pp. 194-208

48. L. Remy, Metall. Trans. A, 1981, Vol. 12A, 387

49. P. Yu. Volosevich et al, Fizika Metal. Metalloved., 1976, 42 (2) 372

50. J.D. Baird, A. Jamieson, J Iron and Steel Inst., 1966, Vol. 204, 793

51. D. Werner, S. Niedzwiez, S.F. Dirnfeld, The British Foundryman, 1970, Oct., pp. 324-326

52. F. Borik, A Study of Strain-Induced Martensite in Metastable Austenitic Manganese Steels, Thesis University of Michigan, 1975, pp. 1-98

53. L.I. Parfenov, G.A. Sorokin, Metalloved. Tern. Obrad., 1973, Vol. 64 (6) 64

54. V.D. Sadovskii, N.N. Syutkina, D.P. Rodionov, L.V. Smirnov, Fizika Metallov. I. Metallovod., 1983, Vol. 56 (3) pp. 618-621

55. A.P. Gulyaev, V.M. Chulkova, Metallovedenie i Termicheskaya Obrab. Metallov., 1974 (4) pp. 49-54

56. C. Gorlier, C. Amzallag, P. Rieu, J.H. Driver, Fatigue, 1984, Vol. 1, pp. 41-48, EMAS

57. D. Rittel, I. Roman, Int. J Fatigue, May 1989, pp. 177-182

58. C. Pichard, M. Truchon, C. Maillard-Salin, Creusot-Loire Centre de Recherches Unieux, 1983, Rapport N° 1598, Dec., 40 p.

59. P. Poyet, H.P. Lieurade, J. Paszko, Bulletin Cercle d'Etudes des Metaux, 1989, Vol. 15 (17) Com. N° 7, pp. 1-9

60. S.A. Katyk, V.L. Volodin, L.B. Zuev, V.P. Rasshchupkin, N.V. Dadotchkin, V.V. Vulf, Izvest. Vhz. Chermaya Metal., 1985 (6) pp. 99-103

61. V.J. Grigorkin, G.V. Korotushenko, L.M. Kutnestsova, Problemy Prochnosti, 1975, Jan. (1) pp. 8-10

62. O. Saeki, D.Y. Kasamatsu, M. Yamaga, H. Hirano, Tech. Bulletin 1008, 1980, pp. 1-10, Kobe Steel Ltd., Kakogawa Works

63. A. Zharov, M. Mikhalev, V.E. Grishina, Metallovedenie i Termicheskaya Obrab. Metallov., 1980 (2) pp. 40-42

64. M.S. Mikhalev, V.P. Baldina, Litienoe Proizvodstvo, 1974 (6) pp. 27-28

65. H.S. Avery, Austenitic Manganese Steel, Metals Handbook, 1958 pp. 834-842

66. H.C. Doepler, J of Metals, 1952, Feb., pp. 166-170

67. M.S. Mikhalev, V.P. Baldina, Izv. Akad. Nauk. S.S.S.R. Metally, 1974, Jan.-Feb. (1)

68. E. Schürmann, M. Hoffman, Giesserei, 1974 (12) pp. 375-383

69. P. Fremunt, K. Stransky, 32e Congrès Inst. Fonderie Varsovie, 1965, 9 Sept., mem. N° 21

70. W. Amende, 1974, Untersuchungen zur Bruchmechanik heterogener Werkstoffe anhand der kontinuierlichen Messung der Rissausbreitung in Manganhartstahl nach Einstellung verschiedener Gefüge, Thesis Doctor-Ingénieur Technischen Universität Berlin

71. H. Berns, Giesserei 64, 9, 1977, June, N° 12, pp. 323-328

72. D.E. Diesburg, Climax Molybdenum Report I.S.J. 836 Ap 1974

73. F. Borik, W.G. Scholz, J of Materials, 1971, Vol. 6 (3) pp. 590-605

74. H.S. Avery, H.J. Chapin, Welding J, 1954, May, Vol. 33, pp. 459-476

75. F. Maratray, T.E. Norman, Rev. Metallurgie, 1961, Vol. 58, pp. 489-506

76. M.S. Mikhalev, A.A. Leitsinger, Liejnoe Proizvodstvo, 1988 (1) 6

77. J.F. Chinella, ASM High Manganese High Nitrogen Austenitic Steels Conf. Proc. 1992 pp. 143-153

78. Yu.S. Rodin, V.A. Kudrin, M.F. Sidorenko, A.S. Morozov, Izv. Viz. Chernaya Metall., 1979 (11) pp. 32-34

79. B.N. Sukhoti, R.V. Kakabadze, Yu.I. Apanasev, V.A. Kudrin, I.Yu. Rodin, 'Stal, 1984 (2) pp. 26-27

80. M. Devaux, C. Gatellier, M. Olette, R. Vasse, 25e Colloque de Metallurgie de Saclay, INSTN, 23-25 June 1982, pp. 111-118

81. C.W. Farrar, AIME 21st Annual Electric Furnace Conf., 1963, 4 Dec., Chicago, Illinois, U.S.A.

82. A. Behr, Metals, 1967, Oct., 53

83. Kos Bernd, Brevet Européen, N°. 841085368, 1984, 19 Oct.

84. C. Gatellier, H. Gaye, Revue de Metallurgie CIT, 1986, Jan., pp. 25-42

85. P. Fremunt, E. Pant, Slevarenstvi, 1967 (8) 306-309, Trad. CTIF N° 3878, pp. 10-68

86. B. Sansby, K. Knaggs, British Foundryman, 1966, 59 (2) pp. 76-88

87. K. Stransky, E. Pant, K. Kostal, Slevarenstvi, 1968 (1) pp. 12-15

88. Manganese Steel, 1956, Pub. Oliver and Boyd

89. P. Asanti, A.F.S. Cast Metals Research J, 1968, March, Vol. 4 (1) pp. 9-15

90. M.I. Kurbatov, B.S. Speranskii, E.F. Parasynk, Russian Casting Production, 1966 (6) 288

91. M. Holzer, The Foundryman, 1980, March, Vol. 83 (3) pp. 135-141

92. J.P. Aymard, M.T. Leger, B. Lageat, Fonderie, 1976, July, 358, pp. 265-273

93. H. Henych, Giesserei, 1973, 26 July, Vol. 60 (15) pp. 453-457

94. H.S. Kim, C.B. Choi, B.S. Chun, S.S. Cho, J of the Korean Inst. of Metals, 1990, Vol. 28 (2) pp. 119-128

95. Y. Kataoka, M. Higuelin, K. Murai, Imono, J Japan Foundrymen's Soc., 1989, Vol. 61 (3) Issn 0021-4396, pp. 183-187

96. M.I. Kurbatov, D.M. Slonin, Metallurg. Gornorud. Prom., 1972, May-June (3) pp. 43-44

97. L.G. Raskolenko, G.I. Yantsen, V.P. Ksenofontov, Yu.D. Novomejskii, A.I. Bychkova, Metalloved. Tern. Obrabot. Metallov., 1973, 7 July (7) pp. 51-53

98. C. Ouchi, Y. Kohsaka (Conf. Proc.) 98th I.S.I.J. Meeting, 1979, Oct., Lecture N° 5910

99. L. Moulin, M. Lacoude, R. Tricot, A. Gueussier, Revue de Metallurgie, 1970, May, pp. 465-472

100. T. Sasaki, K. Watanabe, K. Nohara, Y. Ono (Proc. Conf.) 102nd I.S.I.J. Meeting, 1981, Nov., Lecture N° 1290

101. H.L. Arnson, Climax Molybdenum of Michigan, Report L-212-178,(1977) pp. 1-28

102. K. Stransky, Foundry Trade J., 1965, 25 Feb., pp. 227-235

103. T. Kato, M. Fujikura, K. Ishida, A. Kimura, Y. Takeuchi, N. Kawasaki, S. Yahagi, Trans. I.S.I.J., 1981, 21, pp. 852-861

104. S. Ando, N. Kimata, Science of Machines, 1974, Vol. 2 (26) pp. 34-40

105. A.V. Berezovsky, V.I. Shumyakov, Svar. Proizvod., 1985 (7) 17

106. U.S. Patent N° 4104504

107. S. Kawasaki, Y. Takenchi, T. Kato, Trans. I.S.I.J., 1980, Vol. 20 (8)

108. M. Fujikura, Matériaux et Techniques, 1986, July-Aug., pp. 341-354

109. O. Saeki, Y. Kasamatsu, M. Yamaga, H. Irano, Kobe Steel Ltd., Kakogawa Works, Technical Bulletin 1008, 1980, pp. 1-10

110. Kobe Steel Ltd., Kakogawa Works, 1979, Feb., Report N° A-78B20, pp. 1-45

111. Creusot Loire Industrie, Mise en oeuvre du Creusabro M., pp. 1-2 Commercial leaflet

112. L.I. Parfenov, G.A. Sorokin, Metallov. Tern. Obrab. Metallov., 1971 (5) pp. 43-45

113. S. Krol, Praktische Metallographie, 1981, 18, pp. 75-81

114. H. Berns, H.G. Franke, Steels Research, 1987, 58 (5) pp. 226-230

115. A. Goldberg, Lawrence Livermore N. Lab. Annual Report, 1985, pp. 149-162

116. N.A. Voronova, I.E. Lev, I.Z. Masinson, G.I. Pluzenik, P.T. Hohlov, Structure et Propriétés de la Fonte et de l'Acier, 1967, Vol. XXVI, Ed. Metallurgija, pp. 231-236

117. R.W. Smith, M. Ghoreshy, A. De Monte, W.B.F. Mackay, T.J.N. Smith, M.R. Viner, (Proc. Conf.) Heat Treatment, 1987, 11-15 May, Inst. of Metals, pp. 85-91

118. J.P. Aymard, M.T. Leger, B. Lageat, Fonderie, 1976, July, 358, pp. 265-273

119. D. Werner, S. Niedzwiez, S.F. Dirnfeld, The British Foundryman, 1970, Oct., Vol. 63, pp. 324-326

120. A. Thekston, P.J. Mutton, Matériaux et Techniques, 1977, Dec., pp. 33-36

121. L.S. Malinov, E.Yu. Kharlanova, G.A. Chikalenko, A.B. Gogol, L.L. Mudrik, Steel in USSR, 1986, June, Vol. 16, pp. 296-299

122. S.E. Kondratyuk, G.G. Lutsenko, Metalloved. Tern. Obrab. Met., 1985 (8) pp. 54-56

123. A.I. Uvarov, , P.P. Karpov, T.P. Vasechkina, Metallov. Tern. Obrab. Metallov., 1975 April (4) pp. 60-62

124. N.C. Howells, E.A. Lange, Modern Casting, 1960, May, pp. 115-119

125. Chen Xi Jie, (Proc. Conf.) 1st Int. Steel Foundry Congress, 1985, 11-13 Nov., Chicago, Ill., pp. 209-217

126. V.I. Grigorkin, I.V. Frantsenuyk, I.P. Galkin, A.A. Osetrov, A.T. Chemeris, M.F. Chernenilov, Metall. Tern. Obrab. Metall., 1974, 16, 68

127. G.I. Sil'Man, N.J. Pristuplyuk, M.S. Frol'tsov, Metall. Tern. Obrab. Metall., 1980, 22, 38

128. Ö.A. Atasoy, K. Özbaysal, O.T. Inal, J of Materials Science, 1989, 24, pp. 1393-1398

129. S. Sawa, Bull. of Japan Inst. Metals, 1979, 18, pp. 573-581

130. H. Masumoto, K. Suemune, Y. Yoshimura, H. Inoue, T. Akasawa, T. Ogawa, T. Ohba, S. Hida, T. Harada, Seitetsu Kenkyu, 1982, 309, pp. 14479-14490

131. T. Sasaki, K. Watanabe, K. Nohara, Y. Ono, N. Kindo, S. Sato, Trans. I.S.I.J., 1982, Vol. 22, pp. 1010-1020

132. C. Ouchi, Y. Kohsaka, M. Ueda, Iron and Steels, 1983, 69, pp. 694-702

133. Yu. Chizhenko, A. Khrapov, L. Agryzko, Metalloved. Tern. Obrab. Metallov., 1977, Feb. (2) pp. 56-57

134. R. Wang, F.H. Beek, Metal Progress, 1983, March

135. H. Habashi, I. Galland, Fonderie - Fondeurs d'Aujourd'hui, 1986, Dec., 60, pp. 31-44

136. T. Sasaki, K. Watanabe, K. Nohara, Y. Ono, (Proc. Conf.) 102nd I.S.I.J. Meeting, 1981, Nov., Lecture N° 1290

137. J. Albrecht, Metallurgia, 1983, Vol. 17, pp. 371-374

138. G. Stein, Steel Forging, 1986, STP 903, 237, Philadelphia, ASTM

139. S.K. Banerjee, 1982, Status Report of Fe Mn Al Steel, Ekton, PA, Foot Mineral Co.

140. T. Kato, M. Fujikura, S. Yahagi, K. Ishida, Iron and Steels, 1981, 67, pp. 587-595

141. T. Kato, M. Fujikura, T. Ishida, A. Kimura, Y. Takuchi, Y. Kawasaki, S. Yahagi, Trans. ISIJ, 1981, 21, pp. 852-861

142. T. Sasaki, K. Nohara, N. Kondo, S. Sato, K. Watanabe, K. Ono, I. Chise, Kawasaki Steel Giho, 1981, 13, pp. 381-397

143. T. Kato, M. Fujikura, N. Kawasaki, K. Ishida, Iron and Steels, 1979, 65, 372

144. R. Niwa, Metals in Engineering, Japan, Vol. 19, pp. 58-59 see also (19)

145. Stahl und Eisen, 1982, 102 (23) 11

146. V. Thien, J. Ewald, W. Voss, Scanning Electron Microscopy, SEM Inc., AMF O'Hare (Chicago), 1983, II, pp. 577-584

147. M.I. Kurbatov, A.S. Nosenko, Ya. Protsenko, E.G. Zemka, Metal Science and Heat Treatment, 1990, Vol. 32 (9-10) pp. 711-713

148. B.K. Zuidema, D.K. Subramanyan, W.C. Leslie, Metallurgical Trans. A, 1987, Sept., Vol. 18A, pp. 1629-1639

149. H.J. Lai, C.M. Wan, J of Materials Science, 1989, 24, pp. 2449-2453

150. A. Shulte, A.I. Korneichuk, A.A. Sherstyuk, V.S. Spernsky, Metallurg. H. Gomory Prom., 1971, Vol. 12 (2) pp. 48-50

151. B.B. Vinokur, C.E. Kondriatyuk, G.G. Lutsenko, O.G. Kasatkin, Izv. Akad. Nauk., SSSR Metal, 1986 (2) pp. 123-127

152. A.A. Sherstyuk, M.I. Kurbatov, B.F. Tumanskii, Metallovedenie i Termicheskaya Obrab. Metallov., 1970, March (3) 48-49, Trans. UDC 669. 15' 74 - 194 : 620-17

153. N.A. Svidunovich, A.I. Garost, Liteinoe Prvizvod, 1985 (5) pp. 12-13

154. A.A. Sherstyuk, M.I. Kurbatov, B.F. Tumanskii, Metal Sci. Heat Treat., 1970, March-April (3-4) pp. 232-233

155. K. Bernd, European Patent N° 0136433, 19 July 1984

156. Y. Kataoka, M. Higuchi, K. Murai, IMONO (J Japan Foundrymen's Soc.) 1989, Vol. 61 (3) Issn 0021-4396, 183

157. V.I. Wlasow, Kosudora, Metalurgizdat Moscow, 1963 see also (93)

158. Z. Krolikowski, W. Ferus, Zeztyty Naukove Politechniki Gdanskiej, 1974 (208) Mechanika XIX, pp. 15-27

159. B.B. Vinokur, O.G. Kasatkin, S.E. Kondratyuk, Izv. Akad. Nauk. SSSR Metal, 1990 (2) pp. 81-85

160. British Railways Board, French Patent Application N° Pub. 2.167 - 743, Eur. N° 730030, 5 Jan. 1973

161. I. Schmidt, J Mater. Science Letters 5, 1986, pp. 475-477

162. H. Gabolde, J. Rieu, Ste. Industrie Minerale, 4e Colloque Int. Abras. N° 3, Grenoble, 1979, 9-11 May, pp. 1-12

163. H.S. Avery, Materials for the Mining Industry, Vail, Colorado, 1974, 30-31 July, pp. 43-77, Climax Molybdenum Co. of Michigan Ed.

164. M.S. Mikhalev, V.S. Dmitrieva, A.I. Zarov, Izv. Akad. Nauk. SSSR Metal., 1975, May-June (3) pp. 136-139

165. Yu. G. Matveev, V.S. Bukhtin, D.I. Tarasko, Metallovedenie i Termicheskaya Obrab. Metallov., 1951, March (3) pp. 51-53

166. A. Oda, H. Miyagawa, Zairyo J Soc. Mat. Sci., Japan, 1985, Vol. 34 (380) pp. 525-531

167. R. Blickensderfer, J.H. Tylazak, Wear, 1983, 84, pp. 361-373

168. T. Fukui, J Soc. Mat. Sci., Japan, 1988, 37 (413) pp. 191-197

169. T. Fukui, Y. Waku, N. Nishimoto, J Soc. Mat. Sci. Japan, 1987, June, 36 (405) pp. 584-589

170. A. Oda, J Soc. Mat. Sci. Japan, 1985, 34 (380) pp. 595-601

171. H.M. Bauschke, E. Hornbogen, K.H. Zum Gar, Z. Metallk. Bd., 1981, 72, pp. 1-12

172. I. Schmidt, A. Gücke, Materials Technology, Steel Research, 1986, 57 (6) pp. 286-291

173. R.C.D. Richardson, Wear, 1967, 10, pp. 353-382

174. A.A. Torrance, Wear, 1980, 63, pp. 359-370

175. R.C.D. Richardson, Wear, 1967, 10, pp. 291-309

176. H.S. Avery, Wear, 1961, 4, pp. 427-449

177. J. Lot, F. Maratray, Bulletin SOFRAB, 1984, May (1) pp. 3-11

178. V.P. Ksenofontov, Metallovedenie i Termicheskaya Obrab. Metallov., 1967 (9) pp. 73-74

179. F. Borick, D.L. Sponseller, J of Materials, 1971, Vol. 6 (3) pp. 576-589

180. D.E. Diesburg, F. Borik, Materials for the Mining Industry, Vail, Colorado,1974, 30-31 July, pp. 15-41, Climax Molybdenum Co. of Michigan Ed.

181. H.L. Arnson, Climax Molybdenum Co. of Michigan, Lab. Report L 212, 160, L 212, 178, pp. 1-27 (1977)

182. B.B. Vinokur, C.E. Kondratyuk, G.G. Lutsenko, O.G. Kasatkin, Izv. Akad. Nauk. SSSR Metal., 1986 (2) pp. 123-127

183. D.D. Howat, D.P. Enright, Fourth Int. Abrasion Colloquium, Grenoble, 1979, 9-11 May (21) pp. 1-29

184. B.B. Vinokur, C.E. Kondratyuk, G.G. Lutsenko, O.G. Kasatkin, Teknol. Organ. Proiz., 1985 (2) 38-40

185. J. Vial, R. de Feyter, E. Morlet, Rev. Ind. Minerale, Special N°, 1970, 15 Feb., pp. 159-167

186. E.W. Atabekov, I.A. Montgait, Metallurg. Cornorud. Prom., 1974, Nov.-Dec. (6) pp. 31-32

187. T.E. Norman, E.R. Hall, ASTM, STP 446 (1969) pp. 91-113

188. T.E. Norman, D.V. Doane, A. Salomon, AFS Trans. 1960, Vol. 68, pp. 287-300

189. J.E. Bevan, Lab. Report J.4184, 1976, 3 Sept., 1-4, Climax Molybdenum Co. of Michigan

190. J.G. Brisson, Contribution à l'Etude des Aciers de Construction Résistant à l'Usure par Abrasion, Thesis Conservatoire des Arts et Métiers, 1976, 25 April, pp. 1-84

191. R.L. Woodward, J Austral. Inst. Metals, 1977, Vol. 22 (3-4) pp. 167-170

192. CETIM - Senlis - France, Bibliographical survey for IMI, April 1994, not published